On the (Im)Possibility of Business Ethics

Issues in Business Ethics

VOLUME 37

For further volumes:
http://www.springer.com/series/6077

Minka Woermann

On the (Im)Possibility of Business Ethics

Critical Complexity, Deconstruction, and Implications for Understanding the Ethics of Business

Springer

Minka Woermann
Department of Philosophy
and Centre for Applied Ethics
Stellenbosch University
Matieland, South Africa

ISSN 0925-6733
ISBN 978-94-007-5130-9 ISBN 978-94-007-5131-6 (eBook)
DOI 10.1007/978-94-007-5131-6
Springer Dordrecht Heidelberg New York London

Library of Congress Control Number: 2012949686

Printed on acid-free paper

Springer is part of Springer Science+Business Media (www.springer.com)

In memory of, and with thanks to,

Professor Paul Cilliers (1956–2011)

who taught me that being 'free of the quiet desperation of clinging too tightly to answers, may be our greatest act of faith' (William O'Daly for Pablo Neruda).

Preface

As a result of my experiences in teaching business ethics from a philosophical perspective to accounting and business students on both under and post-graduate level, I have become increasingly sensitive to the challenges and deficiencies that I believe characterise certain current and popular orientations to business ethics. Specifically, I have found that the traditional orientations to business ethics cannot sufficiently account for the complex dynamics in operation in the world today, including the business world.

Since these orientations are in part based on a number of (primarily modernist) ethical theories, these ethical theories should be subjected to critical scrutiny, and the normative basis of business ethics should be reconsidered. All-too-often, it is assumed that definitions of what constitute the good can be easily determined by referring to pre-established categories (such as the promotion of happiness or the promotion of rational action); or, it is assumed that, by following certain procedural guidelines, we will come to the 'right' answers. Such assumptions can only be supported on the grounds of certain reductive tendencies, whereby, for example, it is postulated that one's experiences or context have no effects on one's views of right or wrong, and that the past necessarily resembles the future.

In business ethics, these assumptions give rise to rule, procedural or compliance-based models of ethics, which are employed in order to engage with the ethical challenges in the workplace, and in the wider business community. Due to the prescriptive nature of these models, the primary task of the ethicist becomes to motivate people to accept, internalise, and act according to predetermined ethical standards. Applying these approaches therefore typically assumes a top-down trajectory; and, at the most extreme, the moral agent bears little individual accountability for her actions besides complying with, and applying, predetermined rules. A primary reason for the popularity of this orientation to business ethics is precisely *because* compliance or rule-based models are easy to codify and enforce in a business setting.

Approaches to ethics that account for complexity are difficult, if not impossible, to systematise, since complexity implies a serious engagement with contingency. Yet, if – as I believe it should be – a primary goal of business ethics is to provide business students and practitioners with sense-making tools and tools of analysis, in order to assist them in ethical analysis and decision-making in the workplace, then the tools that we employ should help us to engage with the complexities of our practices.

To this end, it is useful to search for other ideas concerning the nature of ethics – both in the field of philosophy, as well as further afield. This trans-disciplinary exploration is undertaken in this study, in an attempt to think together the paradigms of postmodernism, complexity theory, deconstruction, and business ethics, so as to challenge the underlying assumptions that inform business ethics theories and practices, and to suggest an alternative conception of business ethics. A number of influential thinkers have guided me on this journey, especially to the extent that their own work is based on the understanding that social systems and organisations are complex. In terms of the complexity literature, the ideas of the philosophers and complexity theorists, Edgar Morin and Paul Cilliers, have proven particularly helpful in developing both an understanding of the features of complex systems, and of how we – as embedded, embodied, and limited agents – can successfully and ethically engage with the complexity generated by our contexts. In terms of the philosophical literature, Jacques Derrida's deep concern with the problems related to the naturalisation of conceptual paradigms, and the negative consequences that arise when we view our reductive models as corresponding with reality, has been instrumental in developing a complex ethical position. Furthermore, through employing Derrida's ideas, it was also possible to present a radical challenge to business ethics, and to pose critical questions concerning the categories of business ethics; questions that may – with time – lead to a broader, richer, and more productive interpretation of business ethics. In other words, and as so beautifully described by Alain Badiou, Derrida's ideas challenged me to 'unclose closed matters'. This study is also not the first attempt at re-inscribing the discipline of business ethics in a broader context. To this end, the work of business ethicists and theorists such as Jane Collier, Andrew Crane, Rafael Esteban, Campbell Jones, Hugo Letiche, Michael Lissack, Dirk Matten, Mollie Painter-Morland, and Mary Uhl-Bien has also been extremely helpful in developing this broader, more complex understanding of business ethics.

As is implied by the above discussion, my concern in this study is with both the development of a complex notion of ethics and the application of this normative position to the field of business ethics (particularly, corporate social responsibility). The study is therefore structured in two parts. In the first part, I develop the philosophical foundation at the hand of a reading of postmodernism, complexity thinking, and deconstruction. Specifically, a postmodern ethics is introduced in the first chapter. I support an affirmative view of postmodernism, in which value judgements are deemed possible, despite not being universally justifiable. In fact, the description provided of postmodernism in this chapter serves to discredit universalism by drawing attention to the provisional, reflexive, contingent, and emergent nature of meaning and knowledge. On a postmodern reading, it is impossible to defend the rigid fact-value distinction (which also divides the fields of business ethics) because – according to this reading – ethics can neither be grounded in a transparent, objective, and predictable reality, nor be justified by an appeal to *a priori* normative categories (since these categories are contextualised within specific practices). Postmodernism therefore has a distinctive anti-foundational slant, which has been the target of much criticism in business ethics, and which, as mentioned, poses problems for institutionalising ethics (as will be further explained in this chapter). The radical challenge that postmodernism poses to modernist accounts of ethics cannot however be ignored, and, as such, we need to find productive strategies for taking up the postmodern challenge. Ways of conceptualising such strategies are introduced in this chapter. However, in order to truly engage with the challenges posed by a postmodern account of ethics, it is necessary to further develop our understanding of complex phenomena and complex thinking.

In the second chapter, I therefore introduce a general and critical approach to complexity. Within this approach, complex systems are viewed as irreducible. In other words, it is deemed impossible to uncover the laws of complex systems; and, since complex systems cannot be fully modelled, any engagement with complexity necessitates a critical engagement with the limits and status of our knowledge claims. This critical engagement is denoted by the 'ethics of complexity'. Furthermore, the ethics of complexity commits us to a complex view of ethics. This is because complexity is inherent to any ethical engagement, yet ethical frameworks are also models, and, like all models, are limited, exclusionary, and incapable of accounting for the complexity of lived phenomena. However, models are also necessary, since we need to reduce the complexity, in order to make sense of our world. It is therefore argued that the best ethical models are those that draw attention to their own limited status; and, in this vein, the provisional imperative – which is a self-undermining imperative – is introduced as a guide for responsible ethical action.

Complexity thinking provides us with a broad framework for engaging in complex systems, but, in order to develop these insights into a robust normative position, the complexity framework should be supplemented with a position that can provide it with philosophical depth. To achieve this end, Derrida's ideas are introduced in the third chapter, where deconstruction is explained and the ethics of deconstruction is explored. It is specifically argued that Derrida's philosophy offers a productive reading of a complex notion of ethics, for the following three reasons: Firstly, Derrida's work on quasi-transcendental or limit concepts provides a means for addressing the methodological complexity of thinking together a system and its environment, and thereby both constitutes as engagement with the ethics of complexity and serves to articulate the ethical interruption of ontological closure. Secondly, in deconstructing the conceptual models that inform our practices, Derrida is able to de-naturalise these models and thereby draw attention to both the ethics of complexity and the supplementary complications that pervade all meaning. In so doing, Derrida opens the door to otherness and difference, despite his radically immanent and contextualised understanding of ethics. Thirdly, in explicitly addressing the ethical-political implications that arise due to the limitations of our knowledge, Derrida's *oeuvre* allows for a sophisticated and thorough analysis of the implications that critical complexity thinking hold in the human domain. Derrida's work therefore lends philosophical grounding to the insights gleaned from the analysis on critical complexity thinking.

In the fourth chapter, which serves as the conclusion to the first part of the study, three operations are introduced, which help us to practically engage with the demands of the deconstructive and provisional account of ethics developed in this study, and which also serve to draw attention to 'the logic of complexity' that contaminates all conceptual schemas. These three operations, which I have termed 'virtues' for a complex world, are: transgressivity (which prevents us from simply reinforcing that which is current), irony (which allows us to recognise and engage with the limitations of a binary logic), and moral imagination (which allows us to successfully engage in critical meaning-producing processes that takes place within specific contexts defined by power and politics). The arguments put forward in this chapter are illustrated at the hand of examples from the organisational context, and from the management sciences and business ethics literature.

Having developed the theoretical foundation, I attempt to illustrate the practical applicability of this foundation in Part II of this study. In Chap. 5, the implications that a deconstructive and complex ethics hold for our understanding of corporate social responsibility (CSR) is investigated. The chapter commences with an overview of the traditional characterisation of CSR, in order to demonstrate that this characterisation is informed by an equalising or commutative understanding of justice, i.e. repaying good with good. On this interpretation, CSR policies articulate the content of the social contract, which is premised on a commutative understanding of just relations between societal and economic interest groups. Derrida however offers a much more radical view of responsibility, one that transcends the reciprocal demands and expectations of a circular economy. Responsibility, on his take, becomes an expression of ethical complexity, which means that, in practice, responsible action always pushes the limits of its own expression. However, this understanding of responsibility cannot form the basis of a substantive ethics, and is often criticised for being practically useless. More specifically, critics are concerned that, if a Derridean view of ethical relations and responsible action are irreducible, undecidable, and non-subsumptive, then it is not clear on what basis moral judgement can take place (the charge of relativism), or of what value business ethics can be (the charge of irreducibility). Both these charges are addressed in this chapter at the hand of a close reading of Derrida's work, in order to show how these charges can be overcome, but also to illustrate the value that a complex, deconstructive ethics holds for business ethics in general, and CSR in particular.

The deconstructive and complexity-inspired reading given of CSR in Chap. 5 is developed into a theory and model of CSR in Chap. 6. To this end, a critical investigation of the three components of CSR is presented, namely corporate identity, the nature of the relations between corporations and society (including stakeholders), and the nature of corporate responsibility. The model of CSR that is derived from this investigation depicts the different domains of CSR (namely, the environmental, social, legal, and economic domains) as embedded in one another, and differentially related to one another. It is argued that these domains are interlinked in complex ways with the corporation, which makes it impossible to conclusively define corporate responsibilities. These complexities also frustrate our attempts at managing our CSR obligations, as our current analytic tools are inadequate in dealing with these complexities. A number of management tools that can help in this regard are therefore also introduced, as is a discussion on the leadership approach and responsibilities that support the theory and model of CSR developed here.

The reconfiguration of CSR is illustrative of a larger challenge that complexity thinking holds for the discipline of business ethics, and, I believe, should form the basis of future research. However, a last issue that is explicitly addressed is the implications that this study holds for teaching business ethics. If business ethics is to play a larger role in our business practices, then it is essential that future business practitioners and leaders develop an understanding of the importance of business ethics. Therefore, in the concluding chapter, an overview of the core components that should be included in business ethics education is given, in order to both deal with the fundamental problems that characterise our times, as well as to promote the future viability of business ethics. The analysis of these core components is based on the complexity and Derridean insights presented in the foregoing chapters, and the aim of the analysis is to unpack teaching strategies that can equip students with the sense-making tools and tools of analysis needed to reflect upon the normative dimensions of complex business challenges. Since these challenges are context-dependent, the analysis does not provide examples of specific pedagogical interventions, because these interventions must be forged within specific environments. The suggestions made in this chapter therefore stand prior to any particular teaching guide or curriculum.

For business ethics to be a viable discipline that can positively influence the business world, business ethics – as a subject and as a practice – cannot be restricted to merely repeating and applying the moral precedents established in the history of philosophical ethics. I believe that we, as business ethicists, have a duty to attempt to extend the scope of business ethics, in order to help business students and practitioners to develop the multidimensionality of thought needed to successfully and ethically deal with the complexities that define our age. This study constitutes a meta-position that can aid in this regard, by highlighting the type of considerations that should inform our ethical engagements in our complex contexts.

Stellenbosch, South Africa
December 2011

Minka Woermann

Acknowledgements

Apart from my indebtedness to Paul Cilliers, I would also like to extend my thanks to my other colleagues from the Philosophy Department, namely Tanya de Villiers-Botha, Louise du Toit, Johan Hattingh, Dirk Louw, Vasti Roodt, Hannes Smit, Liesl van Kerwel, Anton van Niekerk, and my assistant Peter Kruger. It is an honour to work with you. To my parents, Mari and Stefan Vrba; my brother, Sep; my husband Sven; and my daughter Mila: thank you for your love and support.

I would like to thank the *South African Journal of Philosophy* for permission to reprint material from the article entitled 'The ethics of complexity and the complexity of ethics' (co-authored with Paul Cilliers). This material was used in Chaps. 2 and 4. I would also like to thank the *Developmental Leadership Program* for permission to reprint material from the research paper entitled 'Institutions of integrity and the integrity of institutions: integrity and ethics in the politics of developmental leadership' (co-authored with Eduard Grebe), in which an earlier version of the analysis on leadership theories and responsibilities presented at the end of Chap. 6 appears.

Contents

Part I Theoretical Foundation

Part I
Theoretical Foundation

Chapter 1
Towards a Postmodern Understanding of Business Ethics

Abstract In this chapter, a postmodern ethics is introduced as an alternative to the primarily modernist normative basis of business ethics. The study supports an affirmative view of postmodernism, in which value judgements are deemed possible, despite not being universally justifiable. In fact, the description provided of postmodernism in this chapter serves to discredit universalism by drawing attention to the provisional, reflexive, contingent, and emergent nature of meaning and knowledge. On a postmodern reading, it is impossible to defend the rigid fact-value distinction (which also divides the fields of business ethics) because – according to this reading – ethics can neither be grounded in a transparent, objective and predictable reality, nor be justified by an appeal to *a priori* normative categories (since these categories are contextualised within specific practices). Postmodernism therefore has a distinctive anti-foundational slant, which has been the target of much criticism in business ethics, and which poses problems for institutionalising ethics as is explained in this chapter. The radical challenge that postmodernism poses to modernist accounts of ethics cannot however be ignored, and, as such, we need to find productive strategies for taking up the postmodern challenge. Introductory remarks concerning the conceptualisation of such strategies are also presented in this chapter.

Introduction

One of the defining characteristics of the post-Enron world in which we live is the increased focus on business ethics. Today, there are few MBA programmes that do not include ethical and governance issues as part of their curricula, many professional bodies have mandated courses in business ethics, and good corporate governance and sustainable business practices are central to what is viewed as successful business. At face value, the focus on business ethics seems to represent a positive development: one that can potentially restore public trust in business dealings.

M. Woermann, *On the (Im)Possibility of Business Ethics*, Issues in Business Ethics 37, DOI 10.1007/978-94-007-5131-6_1,

However, three quarters of a decade after Enron filed for bankruptcy, the world experienced the largest financial crisis since the Great Depression. The world is still reeling from the aftermath of the crisis and the near-global recession that followed, and although there is much debate about what caused the crisis, how best to recover from the crisis, and how to avoid a similar future financial disaster, one thing is clear: the recent focus on business ethics did very little to prevent the crisis. Consequently, many are feeling either despondent or cynical about the role that business ethics can play in tempering capitalism.

For those who take a more sober stance on the matter, the relative failure of business ethics is not so surprising. In many instances, the turn to business ethics constituted a knee-jerk reaction to the business and accounting scandals that rocked the business world in the late 1990s and early 2000s. No doubt it was assumed by many that institutionalising ethics would be enough to curb unethical behaviour in the business world. However, such views are often contingent on the belief that institutionalising business ethics offers a guarantee for ethical behaviour. A more fruitful (and realistic!) approach is to gauge the value of business ethics in terms of its ability to provide *sense-making tools* and *tools of analysis* that can aid ethical decision-making in the workplace.

Along these lines, one can argue that one of the most powerful tools at our disposal is ethical theory, since, to echo John Maynard Keynes (1953: 306), '[i]t is ideas, not vested interests, which are dangerous for good and evil.' This is because ideas – in our case, ideas about right and wrong, good and bad, and what deserves respect and what does not – constitute the conceptual paradigms and models that we employ in order to understand the world. As such, this study is dedicated to exploring the value that certain ethical theories can potentially hold for growing the viability of business ethics.

Traditionally, three main streams of ethical theory have been incorporated in the field of business ethics, namely consequentialism, deontology and virtue ethics. In recent years, the normative basis of business ethics has met with increasing criticism – both with regard to the nature of the ethical theories espoused in this normative tale, and the way in which these theories are commonly appropriated in business ethics. Mollie Painter-Morland (2008) succinctly sums up these problems as follows: these theories are unable to sufficiently accommodate both substance and procedure, both the universal and the particular, and the relationship between individuals and groups. Furthermore, moral claims are often portrayed as universal, and procedures are seen as immutable, with the result that morality is denigrated to the rational application of universal principles to practice. Accordingly, Zygmunt Bauman (1993: 68) argues that moral sentiments are tamed and domesticated 'through lodging them safely in a straightjacket of formal (or formalizable) rules'. Given this line of criticism, Martin Parker (1998a: 289) warns that:

> The warm embrace given to ethics – through professorships in business schools, conference papers, journals, course designs, books... – could all too rapidly turn to suffocation unless we allow ethics some room to breathe. Perhaps it would be better to open a window.

In this study, it is specifically argued that we should open a window for ethics so as to let contingent, contextual, and temporal dimensions back in, so that our theories reflect the complexities that characterise our embedded positions in the world, and the business world in particular. In so doing, we should seek out positions that dispel the illusion of a vantage point from which the ethical horizon becomes clear. Although critical complexity theory and deconstruction will later be shown to represent particular instances of such positions, the aim of this chapter is to give a broad overview of the nature and application of postmodern ethics, in order to explore the *general* challenge that postmodern ethics poses for conceptualising the normative basis of business ethics, as well as to elucidate the value that the broad ideas that constitute this philosophical movement hold.

Characterising Postmodernism

Analytic Distinctions

The term 'postmodernism' first appeared in the late 1960s, but was popularised in Charles Jencks's (1975) book, entitled *The Language of Post-Modern Architecture*. In this book, Jencks contrasts the postmodern architectural style, which is typified by asymmetrical and decentred buildings, with modernist buildings, which are designed around a centre (Easthope 2001). This decentred and anti-foundational approach is also promulgated in what is commonly recognised as the most influential theoretical expression of philosophical postmodernism, namely Jean-Francois Lyotard's (1984) *The Postmodern Condition: A Report of Knowledge*. Although the term 'postmodern' refers to a number of developments in different fields, its common usages are found in art history and architecture, philosophy, and in general accounts of contemporary society (Easthope 2001).

Apart from these disciplinary distinctions, a further theoretical distinction that can help one to better appreciate the meaning of the term is between *postmodernism* and *poststructuralism*. According to this categorisation, poststructuralists (such as Jacques Derrida and Michel Foucault) tend to concentrate on the work of language and discourse, whereas postmodernists (such as Jean Baudrillard and Lyotard) 'analyse the contemporary social conditions of an epoch' or 'concentrate on the theoretical milieu which has developed to sustain these conditions as a response, or a variety of responses, to modernism' (Linstead 2004: 3).

Despite acknowledging the need for classifications, Stephen Linstead (2004) warns that the above distinctions are not rigid, and that, despite family resemblances, many differences prevail under the postmodern and post-structural writers. Adding to this difficulty is the fact that most of the key figures that typify these movements (such as Derrida, Foucault and Baudrillard), have renounced the labels given to them, and even the terms that they themselves have coined. This is not a frivolous or rhetorical move, but shows a genuine concern for labelling and 'the deadening effects which reduce the multiplicity, paradox and struggle with ideas to a series of homogenous 'positions'' (4). For this reason, Linstead wearily treats postmodernism as encompassing post-structuralism, and the same cautious categorisation will be used in the context of this study.

A last important distinction that must be drawn before introducing some typically postmodern ideas, concerns one's attitude towards postmodernism. In this regard, Martin Kilduff and Ajay Mehra (1997) distinguish between *sceptical postmodernism*, which is the perspective that all interpretations are equally valid; and *affirmative postmodernism* which works with the assumption that, despite not being able to appeal to universal truths, some interpretations are nonetheless better than others. Whereas sceptical postmodernism leads to a gloomy assessment of the social sciences (typified by the sentiment that 'anything and everything goes'), affirmative postmodernism calls for a much more nuanced and careful assessment of the social sciences. In this latter view, 'it should' – as Derrida (1988: 146) reminds us – 'be possible to invoke rules of competence, criteria of discussion and of consensus, good faith, lucidity, rigor, criticism, and pedagogy'. It is with this view of affirmative postmodernism in mind that we progress with this study.

(Anti)Ideological Distinctions

Apart from these analytic distinctions, postmodernism and post-structuralism can also be understood in terms of that which these movements oppose, respectively the 'grand narratives' of modernism and the structuralist view of language as a relatively stable system of signs capable of yielding predictable responses. On this interpretation, postmodernism constitutes a response to modernist limitations and failed ambitions.[1] One should however be careful of merely judging this response as a causal reaction to modernism, as this would imply a linear unfolding of events. Rather, as Linstead (2004) argues in his reading of Lyotard, the post-modern simultaneously precedes the modern, yet is also something that can only be understood through and after the modern. To clarify: the nature of human experience has always been complex, yet different epochs dealt with this complexity in different ways. Whereas pre-modern societies responded through myth, magic, and religion; modern societies viewed the world as being guided by rationality, logic, order, and scientific objectivity. The belief was that if we could attain knowledge of the inner workings of nature, we would be able to control (and resolve) this complexity. Postmodernists, on the other hand, embrace this complexity; and, through historical analyses, reveal the representational inadequacies upon which

[1] Unless otherwise stated, a systematic or comprehensive view of modernism is referred to in this chapter, which is a position that supports totalising theories. This position can be contrasted with critical modernism (best exemplified in the works of Jürgen Habermas), which presents a more contextually-sensitive position – one which recognises a radical critique of reason (associated with many postmodern philosophies), but which tries to avoid this critique through an appeal to an inter-subjective paradigm of communicative action. Such a paradigm is premised on a common-sense understanding of ordinary language, which can help us to recognise a certain sphere of validity through argumentation, and which works with the regulative ideal that a horizon of inter-subjective truth can be assessed.

modernist assumptions are based. Postmodernists therefore attempt to uncover the postmodern within the structures of modernity itself. In this regard, Linstead (5) writes that 'the roots of postmodernism lie alongside the roots of modernism.' Or, otherwise stated, whereas the modernist movement focuses on the centre, the postmodernist movement focuses on the margins. The postmodern is therefore the 'para-modern' or 'paratheory', which:

> looks for the fissures in [the] cosy state of [modernist] affairs, the failures, the imminences, the bursts of energy, the collapses, the silences and the refusals of the unsaid and the non-known to become the said or the known (5).

According to Linstead, this argument clearly debunks the common accusations that postmodernism has no history. Postmodernists think history differently, in order to highlight the ways in which the past constitutes the present. Far from proclaiming an anti-modernist intellectualism, affirmative postmodernists engage with tradition so as to undermine the solid foundation and great Truths of modernism, and to deconstruct institutional, social, and disciplinary boundaries, in order to give voice to the marginalised and unrepresented 'outside'. Postmodernism can therefore also be seen as a political movement, which seeks to destabilise powerful entrenched interests, and, thereby, challenge the status quo (Kilduff and Mehra 1997). In fact, Stuart Sim (2001: 4) notes that a common feature shared by most postmodernists is 'an almost reflexive dislike of authority', accompanied by a gesture of scepticism and an anti-foundational bias.

Apart from these three features, many postmodernists also share other basic ideas, including an interest in the crisis of representation, the provisionality of meaning, reflexivity, and the decentring of the subject, each of which will be dealt with briefly below.[2]

The Crisis of Representation

The crisis of representation refers to the postmodernist view that objectivity, freed from the biases of language, is impossible. Pure thought cannot be represented in a continuous and unproblematic fashion in speech (an ideal designated by the term 'phonocentricism'). Rather than being simply representative, thought is constitutive. Otherwise stated, thinking is about living in different ways – not about gaining access to reality (Dillon 2000). Since a theory-neutral observation language is impossible, that which counts as facts or truths are – according to Friedrich

[2] It must be noted that these four ideas certainly do not exhaust the themes commonly covered in the field of postmodernism as such. A more comprehensive discussion would also need to reference Gilles Deleuze's eclectic philosophical tomes or his work on the radical relationality of bodies; Foucault's treatment of power relations; Henri-Louise Bergson's work on time, intuition, creativity and virtuality; Georges Bataille's work on desire and transgression; and, the discussions on post-feminism by Julia Kristeva and Judith Butler (to mention but a few other influential postmodern philosophers).

Nietzsche (1995: 92) – 'illusions of which one has forgotten that they are illusions.' In other words, we are shaped by our linguistic communities and intellectual traditions, and truth resides in these social conventions, as opposed to being an *a priori* given. What this implies is that truth is not universal; rather it is provisional, and has the character of a 'little narrative' (*petit récit*) (Lyotard 1984). These little narratives can be viewed as language games, embodying a particular set of knowledge criteria. Knowledge is embedded in a number of diverse discourses, each with its own rules and structures. There exists cooperation and competition within and between language games, and no single discourse is absolutely privileged. John Hassard (1993: 10) argues that, on this view, professional self-justification – or doing science – involves a form of 'serious play' through which we make value judgements on a case-by-case basis, and through which we 'seek to oppose the moves and positions of other players while advancing our own positions'. Knowledge and texts 'represent a series of choices concerning how arguments should be presented, and these choices are embodied in the text' (Kilduff and Mehra 1997: 465). In this regard, the work of the postmodern writer is to violate the norms and destabilise the language games, in order to show that knowledge is not governed by pre-established rules (Lyotard 1984).

The Provisionality of Meaning

Nowhere is the insight regarding the provisionality of meaning and knowledge more prevalent than in the writings of Derrida. In *Of Grammatology*, Derrida (1976) builds on Ferdinand de Saussure's (1960) understanding of language as a system of signs. Contrary to the representational view of language, which holds that words constitute positive entities, Saussure (120) shows how meaning is premised on the differences between signs: 'in the linguistic system there are only differences, *without positive terms*'. Derrida (1976) appropriates this insight, which constitutes a powerful critique of logocentricism, defined as the belief in a perfectly rational language that can (re)present both the world and thought. However, Derrida also deconstructs Saussure's view of language – which is premised on the necessary relation between the signifier (the word or sound-image) and the signified (the concept or meaning expressed by the signifier) – arguing that such a relation affirms logocentricism, in assuming that the signifier communicates some essential meaning (as denoted by the signified). Derrida seeks to show that the signified is itself a signifier. There is no core to signification: all meaning is already mediated through signifiers that are differentiated from one another. Signification is the result of a constant deferral of meaning from one linguistic unit to another; or, as so beautifully expressed in the carnivalesque literature of Mikhail Bakhtin (1984: 202):

> the word is not a material thing but rather the eternally mobile, eternally fickle medium of dialogic interaction. It never gravitates towards a single conscience or a single voice. The life of the word is contained in its transfer from one mouth to another, from one context to

another context, from one social collective to another, from one generation to another generation. In this process the word does not forget its own path and cannot completely free itself from the power of these concrete contexts into which it has entered.

This again affirms the idea that, contrary to modernist beliefs, the world is not transparent to us; our ideas and theories do not capture reality in a natural, intuitive way. However, Derrida (as with Lyotard) allows for the relative stability of contexts of interaction. Derrida (1988) writes that, in putting into doubt the possibility of the values and norms that guide our perceptions of truth, he is not denying that we should submit to the norms of pragmatically-determined contexts, only that we should account for the perceived stability of such practices (a stability which is inevitably characterised by historicity, non-naturalness, ethics, politics and institutionality).

Reflexivity

The previous point captures something of the postmodern attitude of reflexivity. The postmodern approach to knowledge means that we should be critical, and moreover self-critical of our beliefs, since these beliefs are always the product of a limited and exclusionary intellectual tradition, yet serve to constitute knowledge and meaning. Postmodernists rely on various strategies in order to draw attention to the provisionality of meaning; and, in so doing, denaturalise their intellectual positions. Some of these strategies include writing under erasure (which entails placing metaphysical terms in quotation marks or crossing out the terms employed); double-writing (in which a position is simultaneously affirmed and denied, often in parallel paragraphs); and, the use of rhetoric devices such as irony, puns, expressions that are nonsensical in traditional conventions, paradoxes, and performative contradictions (where the content of a proposition contradicts the non-contingent presuppositions that make such a proposition possible, such as in the statement, 'this sentence is false'). Not only does this reflexive view characterise the postmodern approach to knowledge, but also of how postmodernists view humans and human agency.

The Decentring of the Subject

The modernist view of human agency is founded upon the idea of an *a priori* core self, capable of undertaking rational and causal actions, and thereby of being in control of itself and of the environment. This view also constitutes a form of logocentricism, since it presumes that we are fully transparent to ourselves. In the postmodern tradition, this stable, monolithic self is replaced by the 'image of a dynamic, differentiated self', where the structure of self-concept is viewed as 'complex and multifaceted and the process of self-conception as active and fluid'

(Seabright and Kurke 1997: 99). In other words, the subject is decentred in that agency is both the outcome of personal traits and contingent considerations (including considerations of contextually-defined meaning and knowledge). There is no pure self, which means that the challenge lies in creating (as opposed to merely realising) the self. Such a process of self-creation always takes place in space (we are constituted within a network of relations with others) and time ('being' is an emergent process of becoming). The postmodern structure of self-concept is aptly described in the anthropologist Helmuth Plessner's (1972) dictum 'I am, but do not possess myself'; whereas the postmodern process of self-conception is captured by Ernst Bloch's (1969) reply: 'That is why we must become.'

The anti-foundational bias that characterises postmodern thought is clear from the above introduction. This anti-foundationalism has been the target of criticism by applied ethicists, and in this regard it is useful to review some of the poignant meta-arguments in the business ethics literature pertaining to the desirability of developing business ethics on the theoretical basis of postmodernism.

Assessing the Viability of a Postmodern Business Ethics

Although a number of articles on the relevance of postmodern philosophy for business ethics have appeared in the extant literature in recent times (see, for example, the special issues on Emmanuel Levinas and Derrida which appeared in *Business Ethics: A European Review* in 2007 and 2010 respectively), and although many theorists have sought to develop a critique of, and offer alternatives to, the standard normative theories which currently informs business ethics (see, for example, Painter-Morland 2008; Jones et al. 2005; Verstraeten 2000), little attention has been given to undertaking a meta-examination of the value of postmodernism for business ethics as such. A special issue on postmodernism and business ethics, which appeared in *Business Ethics Quarterly* in 1993, remains the key theoretical anchor-point for this debate. Of the articles published in this special issue, Clarence Walton's contribution (entitled, 'Business ethics and postmodernism: a dangerous dalliance') is specifically noteworthy, as is Andrew Gustfson's response to Walton in a follow-up article published 7 years later, with the title 'Making sense of postmodern business ethics'. It is with reference to these two articles that the value of incorporating postmodern ideas into business ethics is explored.

Walton's Scepticism

Walton (1993) understands the aims of postmodernism as follows: to unfix boundaries in order to reveal forms of domination and authority, to provide an umbrella for diverse ideologies, to reorder the present through a re-reading of history, and to critique Cartesian theorists. Although Walton provides a list of what he regards as the strengths of postmodernism (including the postmodern concern with second-order principles such as job rights and employee privacy, an insistence on ethical pluralism and respect for diversity, a greater focus on contextual issues, a deconstructive methodology that aids ethical decision-making in a complex world, a support of minority voices, and a general critical disposition towards the status quo), he nonetheless believes that postmodernism 'is more threat than promise to the healthy development of business ethics' (286). He provides an extensive list of what he perceives to be the weaknesses of postmodernism, and also identifies a number of primary challenges that will limit the success of a postmodern business ethics. In this regard, postmodernism's lack of conceptual clarification, its reliance on performative reflexivity, and its lack of a solid foundation deserve attention.

The first challenge was already encountered above in attempting to provide an adequate definition of postmodernism, and boils down to the problem of defining that which resists definition. In this regard, Walton cites Fredric Jameson (1991: 342) – a major American analyst of postmodernism – who states that postmodernists' 'effort of conceptual unification is, to say the least, strikingly inconsistent with the spirit of postmodernism itself', but also adds that Jameson himself later recognised that a theory can 'embrace diverse strands and yet remain a coherent theory' (Walton 1993: 287). However, for Walton, the point remains that the *lack of conceptual clarification* regarding both the definition and the content of postmodernism may stretch the limits of tolerance to an unacceptable point, in that it 'fails to present alternatives for what it destroys and does not provide a basis of thinking' (298).

This criticism leads to the second challenge associated with postmodernism, namely that it revels in a paradox (which manifests as a *performative reflexivity*): postmodernism 'does not abandon modernist notions despite its opposition to them. It allows for a description of its essential identity at the same time that it undermines such a rigorous claim to truth'. This is reminiscent of Habermas' (1987) criticism of a radical critique of reason; and, according to Walton, means that it becomes impossible to conclusively justify commonsense prohibitions against rape, murder, and torture – prohibitions, which, under modernism, are founded in unconditional ethical certainty.

The postmodern emphasis on the provisional nature of the moral world leads to the third postmodern challenge, namely that no universal, external standard of ethical behaviour exists. In other words, *no solid base* for business ethics can be established, which – for Walton (1993) – means that, in principle, all moral judgements should be treated as equally valid. Although such a conception preserves individual autonomy, it becomes impossible to act as an authoritative source on ethical issues. Walton (292) contextualises this point by stating that 'while there are solid reasons to worry about financial capitalism and global corporations, their pathologies will not be addressed effectively by an ethics that, at this level of discussion, treats contradictory moral judgments as equally valid.'

Although the business ethicist Martin Parker (1993) seems to have changed tact in recent years (as will become clearer later on in this chapter), he too has spoken out against postmodernism, arguing that without standards of evaluation, it is not clear in the name of what we are criticising. For the 'earlier' Parker, ethics is essentially about saying that 'I think the world would be a better place if such and such were the case' (209). This might mean dealing with contesting ethical-political claims which, according to him, postmodernists are ill-equipped to do. Parker views the epistemological relativity of postmodernism as an escape route to dealing with ethical and political responsibilities, and as an excuse for leaving the difficult choices up to others. For this reason, Parker (1993) ascribes to a critical modernism, which can allow for 'rigorous analysis of organizational changes within global capitalism' (212).

Gustafson's Defence of Postmodernism

Both Walton and Parker hold a sceptical view of postmodernism, in which it is believed that it is impossible to undertake value judgements, with the result that postmodern tolerance moves ethical discussion dangerously close to moral subjectivism or even indifference. Gustafson (2000), on the other hand, adopts a more affirmative view of postmodernism, arguing that one need not appeal to universal truths in order to explain morality. Furthermore, he points out that postmodernism is not the only position to oppose universal truths. Noteworthy in this regard, is the fact that critical modernism itself draws on local contexts in order to establish a sphere of validity and 'truth'. However, even in the absence of an inter-subjective horizon of truth (as is posited in critical modernism), theories of truth are still possible. In this regard, Gustafson cites Derrida's (1988: 146) description of deconstruction as an attempt to reinscribe values in 'more powerful, larger, more stratified contexts'. Furthermore, as Gustafson points out, postmodern morality can be justified on principle-based grounds, despite the lack of a foundational ethics: postmodernists do accept, employ, and justify the use of certain principles. This is evidenced by Derrida's concern with justice, Levinas' concern for the Other, and Foucault's focus on freedom, to name but three examples. As such, Walton's claim that postmodernism necessarily leads to relativism is unfounded. However, a consequence of the postmodern viewpoint is that, in the absence of universal truths and principles, ethics becomes thoroughly contextualised.

Another weakness identified by Walton that Gustafson (2000) contests, is the idea that the emphasis on tolerance causes ethical indifference. This may be true of certain relativist positions, but many postmodern theorists – such as Levinas or Bauman – are critical of the limits of responsibility, as defined within a modernist perspective, and, instead, argue for 'more responsibility and a louder call to vigilance than any modern theory normally invokes' (649). Gustafson further argues that postmodernists were successful in drawing attention to the interests of the marginalised and have (to a large degree) revitalised this interest 'by bringing modernity to face its own principles and to show the hypocrisy within modernity itself' (649). Otherwise stated, by highlighting the limits of our knowledge and ability to know truth, postmodernists have denaturalised modernist ethical positions. However, Gustafson does concede that, although not necessarily anti-theistic or pagan, postmodernism can – in the very recognition of our limitations – 'bring about a sense of awe and humility, akin to worship' (649).

Gustafson's review of Walton's take on postmodernism certainly highlights the need for a closer appraisal of postmodern ideas. In this regard, postmodernism's lack of conceptual clarification (particularly as concerns competing definitions of truth and the good), its reliance on performative reflexivity (which manifests as a result of a radical critique of reason), and its preference for a contextually-defined (as opposed to a foundational) ethics warrant further attention, in order to ascertain to what extent these traits truly pose a challenge for postmodernism's institutionalisation in a discipline such as business ethics.

Postmodern Insights: Redefining the Agenda for Business Ethics

The Status of Contesting Knowledge Claims

The goal of modernist ethical theories such as consequentialism and deontology is to determine rules or decision-making procedures for right actions in any given case; and, to explicate these rules in a manner that is directly accessible to (and can easily be applied by) everyone. However, the hope of establishing a categorically-binding ethics is plagued with problems, most notable of which is that it rests on an outdated and incorrect modernist assumption regarding the nature of reality (as essentially simple and transparent) and human agency (as essentially causal and rational). No normative theory can substantively exhaust the content of moral precepts, or completely define the rules and procedures that will categorically result in moral decisions and actions. Despite this, we should *actively resist* a subjectivist or relativist position to ethical decision-making, as is characteristic of sceptical postmodernists.

The charge of relativism only works if one assumes that such an all-embracing, categorically-binding principle should exist. In other words, the term 'relativism' only makes sense when contrasted with 'absolutism' or 'universalism'. Everything stands in relation to other things (is relative to other things), except the absolute. The fear of relativism is therefore something that has always haunted absolutist, transcendental, or universal projects, and something that was assumed could be overcome with rigorous scholarship. Bauman (1993: 42) explains as follows:

> However difficult the practicing of moral universality proved to be, no practical difficulty was allowed to cast doubt on universality as an ideal and horizon of history. Relativism was always merely 'current'; its persistence in spite of present efforts tended to be played down as merely a momentary hitch in an otherwise unstoppable movement toward the ideal [namely, '[t]he dream of universality as the ultimate destination of human kind'].

In business ethics we see that the attitude expressed in the above citation is often adopted by the architects of moral codes, who view codes as the way to improve upon human conduct. What such architects strive for are non-ambivalent, non-*aporetic* codes that surely await us in the future. However, if we forgo this dream of universality, or the idea that the perfect code for governing moral behaviour can (and will) be written, then the charge of relativism also loses its sting. This is, in effect, a very liberating move, since the modernist project has always been dogged by radical doubt, which follows it like a shadow that threatens to subsume the light of the ideal, and expose the project in all its brittleness.

But, where does this leave us? In light of affirmative postmodernism, the claim that all ethical positions are relative to one another need not mean that all ethical positions are equally valid or valuable. The philosopher and complexity theorist, Paul Cilliers (2005: 260; 263; my italics), argues that '*[l]imited knowledge is not "any" knowledge*... We can make strong claims, but since these claims are limited, we have to be modest about them.' To make a strong claim means that we should actively resist vagueness in our practices and thought processes. Our statements and our positions should be intelligible, even though they account only for a limited perspective. As Cilliers argues, there can be no excuses for the 'ethical' sloppiness that leads to bad decisions and actions.

The Ethical Task: Learning to Reflect on, and Engage with, Ethical Problems

If all ethical positions are not equally valid, then it is important for the moral agent to learn how to engage in moral thinking and decision-making. Although normative ethical theories have an important pedagogical value, the moral agent should not be seduced into thinking that 'ethical problems can be solved in a quasi-technical way through the application of certain procedures or through the application of formal models of reasoning' (Verstraeten 2000: 3). Normative theories should become *tools* for ethical argumentation and decision-making, not blueprints that can dictate how we should act. This implies that apart from considering moral problems from the perspective of these theories, the moral agent should also learn how to *analyse* and contextualise problems within a larger social setting, before ethically reflecting on potential solutions to such problems.

In organisational terms, this means that one should focus on how ethics is formed and contested in practice, the discourse in which ethics is enacted, and the manner in which ethical subjectivity is formed within organisations (Clegg et al. 2007). One of the most important goals of business ethics should be to teach students how to reflect upon moral problems. Business ethics has an undeniable practical component, and to forget this is to engage in fundamental ethics, which is concerned with the rational justification of moral norms and principles. Reflecting on ethical problems therefore cannot be divorced from the complexities of real-life management practices; and, theoretical and conceptual frameworks should be employed that take account of the ways in which ethics is 'differentially embedded in practices that operate in an active and contextualized manner' (111). On this count, moral codes and models cannot determine ethical practice. Rather – to reiterate the point – '[t]hey become instruments that skilful and knowledgeable [organisational] members can engage and play with freely in their everyday management of their own and others' affairs' (112).

This argument is therefore not against ethical rules. Contrary to the dichotomy that Bauman (1993) sets up between the irrational moral impulse and rational rules, a more fruitful approach is to view rational rules as being in service of ethical decision-making. To this end, René ten Bos (1997: 1001) argues that 'ethical rules can do much more with people than just desensitize or stupefy them.' To reinforce this point, ten Bos refers to Foucault's (1982) distinction between *l'agent moral* (who follows the ethical code to the letter) and *le sujet moral* (who consciously chooses a certain attitude with regard to the code). Business ethics, as a technology, creates moral agents; whereas the view of business ethics espoused in this study is one that promotes the development of moral subjects, who use rules and instruments in order to guide, critique, and legitimise organisational practices.[3]

Performative Reflexivity and Moral Judgement

A subtle implication that emerges from this discussion is that, since ethical theories cannot provide all the answers, we are also bound to run up against the limits of rationality and procedural argumentation. If we endow too much faith in logical argumentation, we risk reducing ethics to a 'happy positivism' (Culler 1983). We should therefore acknowledge the inherent limitations and exclusions that characterise our ethical models, as well as the complexities with which we have to grapple when making ethical decisions.

Moral judgement cannot merely be a question of reasoning; choosing a certain interpretation of the good life, and deciding on a hierarchy of values, is as much a matter of personal conviction and individual moral judgement as it is of reason (Verstraeten 2000). Therefore, it is impossible to speak of ethics without also speaking of personal choice. If there is no pure, universal description of how we should act or be, then we inevitably *have to choose* (to abstain from choosing is also a choice) (Cilliers et al. 2002). This is not a comfortable situation. As Bauman (in Bauman and Tester 2001: 46) states: 'being moral means being bound to make choices under conditions of acute and painful uncertainty'. Being uncertain not only means that we do not know what the best possible course of action is, but also implies that we should acknowledge and grapple with the *aporia* that we encounter the moment our arguments run up against the limits of logic. Often it is exactly these moments that reveal the deepest insights into the nature of ethics, and of being as such.

---

[3] In this study, the terms '(moral) agent' and '(moral) subject' are used interchangeably. However, the terms are employed in the spirit of Foucault's description of the moral subject as someone who consciously chooses a certain attitude with regard to the code.

Such encounters however also open up this characterisation of ethics to the charge of the performative contradiction (defined earlier). As previously stated, the radical critique of reason is vehemently opposed by Habermas (1987), who argues that such a critique results in the decentring of the subject and the de-sublimation of our conception of reason. Nevertheless, as Cilliers (2005) reminds us, if we acknowledge that things are complex, then we should also acknowledge the fact that performative tension is inevitable. He continues in stating that:

> We are playing in what Wood (1990:150) calls the "theatre of difficulty", and this requires a certain "performative reflexivity" (132). We need to demonstrate the difficulties we are in, also in the way we talk about them. Our discourse should reflect the complexities. To talk about the complex world as if it can be understood clearly is a contradiction of another kind, and this is a contradiction with ethical implications... It is only by acknowledging that we are in trouble that we can start grappling with the complexities around us... In Derrida's (2000:467) words: 'There is ethics precisely where I am in a performative powerlessness' (261).

This complex, postmodern characterisation of ethics denotes a *modest position*. However, '[the] modest position is not weak, it is responsible' (261). The modest position – which emphasises both choice and a measure of performative reflexivity – does not however imply that we shouldn't do the necessary calculation and groundwork that precedes any responsible judgement or decision. Rather, the point is that such calculations cannot absolve us from making decisions. In this regard, Derrida (1988: 116) writes that a 'decision can only come into being in the space that exceeds the calculable program that would destroy all responsibility by transforming it into a programmable effect of determinate causes.' Geoffrey Bennington (2000: 15) spells out this insight in terms of ethics: 'Ethics' he writes 'begins where the case does not correspond to any rule, and where the decision has to be taken without subsumption.'

The Ethical Task: Broadening Perspectives on Available Choices

If, as Clegg et al. (2007: 111) rightly maintain, choice does not involve complete free play, but 'proposes an oscillation between possibilities, where these possibilities are determined situationally', then business ethicists should help students and practitioners to develop a more comprehensive understanding of the different choices and decisions that are open to them in any given situation. In order to facilitate this process of reflection, business ethicists should also provide students and practitioners with the means of 'breaking out of their closed or limited hermeneutic circles' (Verstraeten 2000: ix). Although this will not dispel the anxiety that characterises ethical decision-making, it may help to broaden the perspectives of students and practitioners alike, thereby sensitising them to the exclusionary nature of choice (to choose x is to deny non-x), and to gain more robust and diverse perspectives on situations. In order to illustrate this point, Rolland Munro (1998) uses the example of bookkeeping: much of what is deemed important in society is excluded from the structures of bookkeeping, including common goods (such as clean air) and social capital (such as community health). Without a price these societal goods are often devalued.[4]

Johan Verstraeten (2000) maintains that by exposing students and business practitioners to literary and philosophical texts that offer a horizon of interpretation which is different to the scientific or positivist paradigm in which they have been trained, business ethicists can help students and practitioners to consider more variables and make 'better' calculations before undertaking decisions. In terms of developing business ethics pedagogy and its applicability in the business world, Verstraeten advocates a view of business ethics as an 'integral life-enabling education' (vii). The argument for such an education is that:

> actual and future professionals are not sufficiently prepared to deal with the ethical aspects of their professional decisions and with the social consequences of their work. They need a broader education in which their professional knowledge and expertise is completed with the ability to resolve ethical dilemmas and with the capacity to discern the values that are at stake in every professional decision (viii).

Such an education would comprise 'a broadly literary, philosophical, and cultural education that provides future professionals with the capacity to 'meaning-fully' interpret the reality within which they live and act' (x); and, in this regard, promote the true voyage of discovery, which, as Marcel Proust (1934) reminds us, is not about seeking out new territory, but about learning to see with new eyes.

The Contextually-Defined Nature of Ethical Practices

Although the possibility of ethical action rests with the individual moral agent as ethical decision-maker, it is impossible to consider the questions 'what ought I to do?' and 'what type of person should I be?' without accounting for the particular *context* in which the moral agent is embedded. In this regard, Painter-Morland (2008: 87) describes the moral task as follows:

> Moral agents are required to remain fully engaged with the concrete contingencies and dynamics of the world. Instead of an abstract cognitive exercise, ethics as practice is all about participation, relationships and responsiveness.

Our ethical responses and the type of people we become are inseparable in a world where the descriptions that we attribute to ourselves and our actions are never neutral. In the absence of a grand universal scheme, all our decisions and actions take on political and ethical significance (Cilliers et al. 2002). This is because we act, and are acted upon, by each other; and are *constituted* through our practices and through engagement with the world. This process of identity

---

[4] Indeed, triple-bottom-line reporting (Elkington 1999) constitutes an attempt to extend the scope of bookkeeping to account for some of these social goods, most notably social and environmental capital.

formation also helps to shape our conceptions of what it means to be human, and how we should relate with the world (Cilliers and de Villiers 2000). Therefore, contrary to the modernist project, which presents the self as a fixed and coherent individual (accessible to both herself and others), we see that knowledge about the self is limited, contextual, and temporal. On this understanding, ethics is 'one of the many practices in which individuals engage in order to constitute themselves into subjects' (Keleman and Peltonen 2001: 162).

The example of affirmative action programmes adequately illustrates the above point: despite the active propagation of equal employment opportunity (EEO) legislation in America, it has not been sufficient to gain equal status for women in organisations. Drawing from the literature on affirmative action, Clegg et al. (2007: 112) argue that the reason for this is because the legislation does not reach 'the tacit micropractices of everyday organizational life' through which discrimination is enacted. Here we see how organisational stereotypes and prejudices cannot be eradicated completely through legislation, but are – to a large extent – the product of the power and agency of those organisational members who interact to create gender inequality. This example serves to illustrate both the point that ethics is culturally-driven and enacted within a specific context, and that ethics is not defined by *a priori* schemes, but concerns processes of self-formation amongst organisational members. As such, ethics always comes with a history; we are embedded – or 'emplaced' (Munro 1998) – within certain circumstances that serve to structure our ethical paradigms.

The Ethical Task: Nurturing a Critical Disposition

Given the above description, business ethicists should encourage students and professionals to exercise vigilance, and to reflect upon the specific values to which they attach importance, and which inform their frames of meaning. Reflecting on one's values and frames of meaning also necessitates consideration of the tradition to which one belongs, since there is no vantage point from which we can act or judge in a neutral or disinterested fashion:

> Like it or not, one always belongs to a tradition of thought or belief or, in a fragmented culture, to various traditions from which one draws inspiration. Even when one tries *a priori* to put the influence of tradition out of play, one belongs to a tradition, namely the tradition that uses this conception (Verstraeten 2000: xi).

Unlike a de-politcised communitarian response to tradition, business ethicists should help students and practitioners to provide 'a *critical* account of how we came to believe what we do about ourselves and the world' (Painter-Morland 2008: 90; my italics) when we reflect on our frames of meaning. Such frames of meaning cannot be justified on the basis of epistemological or ontological arguments. We should rather offer *ethical-political reasons* for why we judge our ethical schemes and values to be important, whilst simultaneously accepting the exclusions, limitations, and provisionality of such schemes. Furthermore, we should also recognise the fact that, because we are embedded within a specific context, such a critical account is never the outcome of the intentional actions of a free moral agent, but always develops at the hand of a communicative process with others. Painter-Morland (2006) provides the following reasons for this: firstly, the complexity that defines any situation makes it impossible to isolate a single account of an event, or a single cause for change. Secondly, we cannot step out of the network of relations that constitutes our context, in order to exercise judgement. And, thirdly, even if it were possible to do so, any information would immediately feed back into the system and produce a number of new and unpredictable effects in the system. Nevertheless, despite these *caveats*, communicative processes that focus on critical accounts of our contexts provide an important forum for challenging the status quo, and for living vigilantly.

The above constitutes introductory remarks on the type of ethics that is supported by a postmodern disposition. The nature of this broadly postmodern ethic will be fleshed out in subsequent chapters at the hand of specific theories and examples. However, at this point, it should be clear that Walton's argument that the anti-foundational nature of a postmodern business ethics cannot offer business students and practitioners any guidance on their moral problems and dilemmas, rests on a biased and negative view of what constitutes postmodernism. Although postmodernism's anti-foundationalism certainly presents problems for institutionalising ethics, the radical challenge that it poses to modernist accounts of ethics cannot be ignored, and as such it is important to develop productive strategies for dealing with this anti-foundationalism.

Critical Challenges: Problematising the Fact-Value Distinction

An important consequence that arises from the analysis thus far is that – from a postmodern vantage point – the fact-value relation should be viewed in differential rather than oppositional terms. Given this understanding, it is impossible to defend the rigid fact-value distinction (i.e. one cannot derive an 'ought' from an 'is'), which serves as a necessary condition for any foundational account of ethics, and upon which the normative and descriptive fields of business ethics are based. The consequences that this distinction holds for business ethics will be briefly examined, before presenting an argument for why this distinction is untenable.

The Fields of Business Ethics

In their article entitled, 'Business ETHICS / BUSINESS ethics: one field or two?', Linda Trevino and Gary Weaver (1994) argue that academic business ethics is divided into a normative or prescriptive interpretation, and an empirical or descriptive interpretation. Whereas the former interpretation stresses ethics above business, the latter places the emphasis on business above ethics. Although one should be weary of an over-simplification of these interpretations, one could state that – in broad terms – the fissure between these interpretative categories is marked by a difference in meta-theoretical assumptions that guide the theories and norms underscoring these interpretations.

Briefly stated, the normative field is concerned with judging and justifying behaviour as morally right or wrong, whereas discussion within the descriptive field is limited to questions concerning epistemology and methodology, and critical faculties are focused on determining ethical yardsticks (Willmott 1998: 79). According to Parker (1998b), the clash between the normative and descriptive interpretations can be crudely set up as idealism versus realism, or – in metaphorical terms – the ivory tower meets the law of the jungle. Below follows a more detailed description of the differences between these two approaches, at the hand of Trevino and Weaver's article.

The Normative Field

The normative approach – which focuses on what *ought* to be the case – is interdisciplinary in nature, drawing from 'philosophy, theology, political and social theory, and other self-consciously critical inquiries' (Trevino and Weaver 1994: 114). Trevino and Weaver concede that the normative task is not only prescriptive in nature, but also involves analysis and description; however, they nevertheless identify the formulation of prescriptive moral judgements as the dominant feature of the normative approach. The normative approach is, therefore, 'unashamedly value-driven' (116).

Although the task of formulating these moral prescriptions is not specified by any research methodologies (as is the case in the social sciences), there is nevertheless a methodological self-consciousness, which individuates to the task at hand, and which can be described 'by a small number of heuristic guidelines' (116). Whilst the philosophical methodology lacks specificity, the philosophical lexicon is well-developed, having been refined over centuries of philosophical and ethical study. For example, the term 'ethical behaviour' refers to behaviour that is right, just or fair – and each concept, in turn, has its own history. In order to determine whether these terms are understood and used correctly, a significant amount of conceptual clarification precedes the actual formulation of moral prescriptions or judgements. According to Trevino and Weaver, this meta-ethical task is exemplified in the debate over the ontological status of organisations (i.e. are organisations moral agents to which we can attribute moral responsibility and blame?).

Apart from this meta-ethical task, the purpose of normative business ethics is to evaluate the propriety of the corporate world, and to prescribe morally-better alternatives. These morally-better alternatives are encapsulated in moral standards or principles, which are derived from normative ethical theories. Standards and principles present abstract, ideal cases, against which real actions can be critiqued and evaluated. In other words, normative business ethicists, in specifying what ought to be the case, are more concerned with the instantiation of a moral principle, than with the causal antecedents of an action.

Opinions differ as to how moral principles should be applied to business problems, but it is generally accepted that normative ethical theories provide us with the tools necessary to undertake an analysis or conduct an informed discussion on ethical issues that arise in the business context. Furthermore, opinions also differ regarding which principles should be applied to resolve business ethics problems. Generally, however, MacIntyre's (1984) circumscription of applied ethics as containing two elements is accepted. These two elements are 'context-neutral (i.e. putatively universal) ethical theory, and context-sensitive discussions of particular ethical issues' (Trevino and Weaver 1994: 121). Decisions regarding which normative theory to apply to a given business ethics issue depend on whether a given theory is more 'correct', useful, or well-founded than contesting theories.

According to Trevino and Weaver (121), the scope of normative ethical theory 'concerns morality as such i.e., a standard of moral reasoning which holds for persons qua persons.' It is assumed that moral persons or agents – who have to make choices in practical contexts – freely and responsibly decide whether to act in accordance with the moral standards espoused in these normative theories. Therefore, because moral agents are rational and autonomous, 'moral action is self-explanatory or self-interpreting in character, needing no additional explanation in causal or nomological terms' (119). The explanation for ethical behaviour lies between the dictates of morality and an agent's actions. Thus, in a nutshell, one can state that the 'method' of ethical theory involves achieving a 'reflective equilibrium between theoretical constructions and our considered moral judgments' (Rawls 1971 in Trevino and Weaver 1994: 122).

The Descriptive Field

In contrast to the normative approach, social scientists who employ the descriptive approach – which Trevino and Weaver (1994) identify most strongly with the functionalist paradigm – attempt to elucidate what is the case. Such a goal is premised on an objective view of the world, as well as a managerial orientation geared towards stability, as opposed to change. In order to define, explain, and predict phenomena in an organisational context, social scientists make use of design criteria and quantitative statistical methods to test the validity of hypotheses. Historical analysis, observation, interviews, surveys, and experiments are also used as methodological tools in the descriptive approach.

Compared to the normative approach, the ethical vocabulary employed by social scientists is quite young. As a result, key ethical terms are used loosely, to refer to different things in different contexts. For example, in the language of the social scientist, 'ethical behaviour' does not necessarily refer to behaviour that is *a priori* characterised as good, right, or deserving of respect, but can represent any behaviour exhibited by individuals facing ethical decisions, or refer to external determinants that impact upon individual behaviour.

According to Trevino and Weaver, most social scientists ascribe to Bandura's (1986) viewpoint of reciprocal causation, where ethical or unethical behaviour is the outcome of both individual and environmental factors, which mutually influence each other. Social scientists, in emphasising multiple determinants of human behaviour, find it difficult to reconcile ethical behaviour with moral free will. Moreover, although conceptions of individual responsibility and autonomy are not negated in this paradigm, external determinants (such as reward systems, the visibility of business ethics tools etc.) are viewed as more interesting subjects of study, as these factors can be controlled and manipulated.

Complex organisational behaviour is described with reference to social scientific theory, which 'provides a conceptual basis for examining regularities and relationships that can lead to generalizations about organizational behavior – to describe, explain or predict specific outcomes of interest to the researcher' (121). Social scientific theory, therefore, forms the theoretical basis for managing the ethical behaviour of individuals and organisations. Hypothesised causal factors are informed by 'the social scientific roots of the investigator' (122) and empirical research is undertaken to determine the strength and influence of causal factors.

The descriptive approach is based on empirical observation, and theory is built incrementally and deductively by testing hypotheses against organisational phenomena and behaviour, in order to determine the explanatory or predictive success of a given theory. The extent to which these theories help managers to deal with, and predict, ethical problems is used as the criterion for evaluating the success of a given theory.

The Postmodern Challenge

Trevino and Weaver argue that the normative and descriptive fields are becoming more entrenched and institutionalised, and that – because the conceptual chasm between these orientations is so vast – it may be better to divide business ethics into two fields. In support of this view, Parker (1998b: 284) argues that the discipline of business ethics is best characterised by Lyotard's term, *agon*, which 'refers to the wrestling match between incommensurable language games'. From a postmodern perspective, this dichotomy (although alive and well in practice), is conceptually untenable, as described below:

Problematising Descriptive Ethics

If knowledge is provisional, our descriptions of the world do not accord with an *a priori* view of reality, but involves certain choices of how to portray reality. This means that all our decisions and actions are characterised by a normative dimension. In this regard, Hugh Wilmott (1998) argues that if our ethical models cannot be justified on epistemological or ontological grounds, then the very idea of descriptive ethics is problematised. In other words, the category of a purely descriptive ethics is only coherent if ethics can provide us with descriptions of that which exists prior to our interpretations, and therefore independently of language. However, as Richard Rorty (1979) reminds us, we do not hold a mirror up to nature. Knowledge is always, already interpreted knowledge; and our interpretations are, by definition, contextual, temporal, and limited constructions. As a result, the category of descriptive ethics collapses into normative ethics, and any attempt at pursuing 'truth for its own sake' always translates into:

> a tacit quest for something more than truth, for other values may have been obscured, denied, and perhaps even forbidden... In this sense, 'truth for its own sake' is a crypto-ethic of concealment of other substantive values (Gouldner 1973: 65).

Willmott (1998) argues that, as ethicists, merely acknowledging the normative component is not enough. We should also engage in analytical ethics, which, more than 'simply voicing an evaluation or judgement of an issue... *prompts reflection* on the basis of such judgement' (80; my italics). More specifically, he maintains that 'the 'analytical ethics' of post-structuralism has a normative thrust as it challenges the coherence of 'descriptive ethics'.' He states that poststructuralist thinking is particularly helpful 'in re-membering the connectedness of what appears to be antimonies – such as 'good' and 'evil', 'fact' and 'value' or 'self' and 'other'' (87).

Problematising Normative Ethics

A second challenge to the rigid dichotomy between the normative and descriptive orientations to business ethics stems from the view that, since normative considerations cannot be viewed in relation to an Ideal realm, they should rather be considered within specific practices and language games. In other words, our decisions and actions should be contextually-defined. Parker's (1998a) later work characterises this view. Parker argues that, if normativity pervades every decision and action, then every element (and description) of organisational life is construed as analytically inseparable from the philosopher's ethics. As a result, ethics becomes thoroughly 'sociologized', as it is drawn from the lofty ivory tower of academia into the flow of the ordinary. Since the ethical cannot claim a special position anywhere, one should 'attempt to go beyond any metaphysics of good and evil and gesture at relativism in the interests of a *thicker description*' (285; my italics).

Instead of providing prescriptive arguments for people's actions, Parker contemplates the merits of engaging more intensely on the descriptive level, in order to see how people live, and how they draw conclusions concerning right and wrong. The drawback to this approach is that such an analysis cannot yield general results, because conceptions of right and wrong vary in, and between, contexts. This leads to what Kilduff and Mehra (1997: 465) term the 'crisis of generalizability'. The outcome of this strategy is that it 'squeezes [ethics] so flat that it gets everywhere, and hence is really nowhere' (Parker 1998a: 294). Decisions and judgements reverberate through endless networks, making it impossible for us to say: here is an instance of ethical action, there is not.

To prevent such a conclusion, Parker suggests that we should embrace the paradox that being ethical may mean giving up on ethics. This strategy presents a more provisional and contextual way of thinking about ethics, as it focuses on the ambivalence of actions and judgements. This, according to Parker, might mean that we need to reassess the manner in which business ethics questions are asked, and possibly argue that such questions are best left unanswered. Lee Cronbach (1986: 91) summarises this sentiment as follows: 'Social science is cumulative, not in possessing ever-more refined answers about fixed questions, but in possessing an ever-richer repertoire of questions.' According to Campbell Jones (2003), Parker's agenda can be explained as an attempt to prioritise questions of axiology, position, ethics, and politics over ontology and epistemology. Such goals resonate well with postmodern thinking.

Implications

The difference between Willmott's and Parker's views illustrates the different focal points that exist in terms of how 'we should describe theorizing and theorize description' (Parker 1998b: 4). The former strategy (*describing theorising*) is represented by Willmott's approach, where emphasis is placed on a close examination of 'texts' in contemporary social sciences, in order to unveil the normative assumptions that are smuggled into seemingly neutral descriptions. The latter strategy (*theorising description*) is more in line with Parker's own attempt at foregrounding empirical contexts (including the local, institutional and historical contexts), in order to develop thicker descriptions of our practices.

Both Willmott's and Parker's arguments hold important implications for the circumscription of business ethics – either as a normative, or as a descriptive field. Firstly, the normative orientation to business ethics is challenged if we reposition ethics in 'the flow of the ordinary'. The reason for this is because our decisions and actions are contextualised, which means that universal moral prescriptions are an inadequate basis for evaluating and critiquing corporate actions. Furthermore, the idea that moral agents make free, informed, and responsible choices on the basis of these principles alone, is hardly credible. This is because the actions of moral agents are influenced and limited by contextual factors, and because moral agents are rationally-bounded i.e. we are limited by the resources at our disposal, by our cognitive capacities, and by the amount of time that we have to make decisions.

Secondly, and with reference to the descriptive approach, it should be clear that moral agents function in complex environments. The idea that we can obtain objective knowledge about the social world, and the idea that the social world is inherently stable and certain – in fact so certain that we can give accurate explanations and predictions of human behaviour – are predicated on a discredited functionalist paradigm. Whilst the strength of the descriptive approach lies in its consideration of the impact of external determinants on human behaviour, the objectivist paradigm within which many social scientists work is undermined by Willmott's argument.

If neither the traditional circumscription of the normative and descriptive orientations to business ethics ring true to our lived experiences, it seems that this classification should be reconsidered. The collapse of the fact-value distinction implies that ethics cannot be divorced from our intellectual orientations and from our lived experiences in the world. It is precisely this imbrication of ethics with the world that introduces *ethical complexity* and, for Neville Wakefield (1990: 151), this means that:

> We find ourselves left with something more modest, but perhaps more urgent... That is the task, not of defining ends, solutions and finalities, but of living in a world from which these privileges and certainties have been withdrawn.

Postmodern Ethics as an Ethics of Practice

Despite the theoretical nature of the above argument, it supports the expectations of business students and practitioners with regard to the value of an ethics education. This is clear from an unpublished survey on improving the teaching of business ethics for accounting students, conducted by Woermann and Hattingh (2008) at three South African universities. The students' responses resonated with the view of business ethics as contributing to 'the enhancement of life' (Painter-Morland 2008: 237). For example, one respondent said that the goal of business ethics is 'to get someone to start thinking critically – it motivates students to ask why we are here... [Business ethics focuses on] more than just money or more than just enjoying life.' Another respondent echoed these sentiments, saying that business ethics is about developing a 'conscious lifestyle'. Yet another respondent mentioned that business ethics 'helps you to think' and 'brings the world of ethics closer to home and lets you look at yourself, [and] critique your own decision...' A number of responses were also contextualised in terms of the accounting profession. Professional reasons for supporting business ethics education included the perception that 'ethics is a very integral part of being an accountant, especially with more reliance on good information in the capitalistic markets today'; and, that 'without due trust in the integrity of our professionals, we have little reason for them'. Given these viewpoints, as well as the theoretical premises of postmodernism, ethics should be understood as something that *constitutes* both our knowledge and us, rather than merely as a normative system that dictates right action

As the title of her book *Business Ethics as Practice: Ethics as the Everyday Business of Business* suggests, Painter-Morland (2008) supports the view that, in order to be effective, ethics should become an integral part of business. When ethics is seen as a part of business practice – rather than as an afterthought to business-as-usual – our understanding of business ethics also shifts from ethics as a mere insurance policy or compliance model, to ethics as a careful (and continuous) consideration of how we wish to live and who we wish to be. Painter-Morland argues that it is ultimately this goal that business ethicists should advance in their work, and she defines 'ethics as practice' as the task of removing

> the conceptual and procedural restraints with which we have tried to secure morality, and continually allow ourselves to be challenged by various contextual and relational realties that fill our everyday lives (93).

When business ethics becomes a practice, self-aware moral subjects have to account for their conceptions of the good (along with the traditions that support such conceptions), as well as how their morals are enacted in practice. On this account, substantive concerns can never be done away with, as – contrary to the modernist conception – we are not autonomous individuals making our own rational and optimal choices. In this regard, Bauman (in Bauman and Tester 2001: 46) writes: 'being moral means knowing that things may be good or bad. But it does not mean knowing, let alone knowing for sure, which things are good and which things are bad.' We have to thus be continuously aware of how our actions both form us, and affect others. Note however, that this stance is not the same as the version of self-surveillance propagated by Kant (2008) (where, as Munro (1998) notes, we must avoid despising or condemning ourselves secretly in our own eyes). We can never have knowledge of the correct decision ahead of the fact, because both the world and moral subjects are complex, and our actions do not unfold linearly and predictably in time. Morality is not abstract, but grounded in the everyday problems of real people living their lives. As such, our decisions are never undertaken in a state of free play, but are limited by the structures, power relations, and networks in which we partake – a point which is also stressed in postmodern philosophy.

Ethics therefore requires both self-awareness and an awareness of the communities in which we live and work. We need to be able to simultaneously deal with the individual and the communal, and the personal and the structural, in order to avoid accounts of human behaviour that presuppose either absolute free will (as is assumed in many ethical theories), or moral determinism (which vacates the space of the moral agent, as is the case in the functionalist paradigm and rational choice models). Furthermore, we should also seek out positions that can accommodate both the normative and the descriptive dimensions of what it means to be human and to live in the world (or, that can accommodate both the ways in which we describe our theories and theorise our descriptions).

If the ethical task is only construed in terms of reflecting on moral judgements, we risk turning ethics into a technology for textual critique, whereby the normative exclusions in discourses are exposed. However, if we view the ethical task as merely producing thicker descriptions of highly-contextual practices, we risk developing a relative and incommensurable view of ethics. Only by simultaneously engaging in textual critique, and investigating the contexts in which our ethical decisions are enacted, is it possible to develop a productive reading of ethics. Such a reading can account for the norms that guide our relatively stable practices, whilst preventing these norms from becoming naturalised. Since Derrida's philosophy will largely inform this study, it is worthwhile to note that – despite typically being associated with the post-structural tradition of textual critique – he is an advocate of this double-handed strategy, and explicitly argues that close reading does not only refer to textual analysis, but also contextual analysis. In this regard he writes:

> I would assume that political, ethical and juridical responsibility requires a task of infinite close reading. I believe this to be the condition of political responsibility: politicians should read. Now to read does not mean to spend nights in the library; to read events, to analyze the situation, to criticize the media, to listen to the rhetoric of demagogues, that's close reading, and it is required more today than ever. So I would urge politicians and citizens to practice close reading in this new sense, and not simply to stay in the library (1999: 67).

It is posited that this view of a postmodern ethics, as an ethics of practice, can benefit the discipline of business ethics, especially to the extent that such a project is geared towards conceptions of ethics that both reflect, and help us to productively engage with, the multiplicities and uncertainties that define the world. A postmodern ethics of practice therefore demands an acknowledgement of ethical complexity, and it is specifically to questions of ethics and complexity that we turn next.

References

Bakhtin, M. 1984. *Problems of Dostoevsky's poetics,* ed. and trans. C. Emerson. Minneapolis: Minnesota University Press.

Bandura, A. 1986. *Social foundations of thought and action: A social cognitive theory*. Englewood Cliffs: Prentice Hall.

Bauman, Z. 1993. *Postmodern ethics*. Oxford: Blackwell.

Bauman, Z., and K. Tester. 2001. *Conversations with Zygmunt Bauman*. Cambridge: Polity Press.

Bennington, G. 2000. *Interrupting Derrida*. London/New York: Routledge.

Bloch, E. 1969. *Spuren: Complete works*, vol. 1. Frankfurt am Main: Suhrkamp.

Cilliers, P. 2005. Complexity, deconstruction and relativism. *Theory, Culture and Society* 22(5): 255–267.

Cilliers, P., and T. de Villiers. 2000. The complex I. In *The political subject*, ed. W. Wheeler, 226–245. London: Lawrence and Wishart.

Cilliers, P., T. de Villiers, and V. Roodt. 2002. The formation of the self. Nietzsche and complexity. *South African Journal of Philosophy* 21(1): 1–18.

Clegg, S., M. Kornberger, and C. Rhodes. 2007. Business ethics as practice. *British Journal of Management* 18: 107–122.

Cronbach, L.J. 1986. Social enquiry by and for earthlings. In *Metatheory in social science: pluralism and subjectivities,* ed. D.W. Fiske and R.A. Shweder, 83–107. Chicago: University of Chicago Press.

Culler, J. 1983. *On deconstruction: theory and criticism after structuralism.* London: Routledge.

Derrida, J. 1976. *Of grammatology*, trans. G. Spivak. Baltimore: Johns Hopkins University Press.

Derrida, J. 1988. Afterword. In *Limited Inc.,* ed. G. Graff and trans. S. Weber, 111–160. Evanston: Northern Western University Press.

Derrida, J. 1999. Hospitality, justice and responsibility: A dialogue with Jacques Derrida. In *Questioning ethics: Contemporary debates in philosophy*, ed. R. Kearney and M. Dooley, 65–68. London/New York: Routledge.

Derrida, J. 2000. Performative powerlessness – A response to Simon Critchley, trans. J. Ingram. *Consellations* 7(4): 466–468.
Dillon, M. 2000. Poststructuralism, complexity and poetics. *Theory, Culture and Society* 17(5): 1–26.
Easthope, A. 2001. Postmodernism and critical and cultural theory. In *The Routledge companion to postmodernism*, ed. S. Simm, 15–27. London: Taylor and Francis.
Elkington, J. 1999. *Cannibals with forks: The triple bottom line of 21st century business.* Minnesota: Capstone.
Foucault, M. 1982. *Histoire de la sexualité 2: L'usage des plaisirs*. Paris: Gallimard.
Gouldner, A. 1973. The sociologist as partisan. In *For sociology: Renewal and critique*, ed. A. Gouldner, 27–68. New York: Basic Books.
Gustafson, A. 2000. Making sense of postmodern business ethics. *Business Ethics Quarterly* 10(3): 645–658.
Habermas, J. 1987. *The philosophical discourse of modernity: Twelve lectures*, trans. F. Lawrence. Cambridge: MIT Press.
Hassard, J. 1993. Postmodernism and organizational analysis: An overview. In *Postmodernism and organizations*, ed. J. Hassard and M. Parker, 1–24. London: Sage.
Jameson, F. 1991. *Post modernism, or, the cultural logic of late capitalism.* Durham: Duke University Press.
Jencks, C. 1975. *The language of post-modern architecture*, 4th ed. London: Academy Editions.
Jones, C. 2003. As if business ethics were possible, 'within such limits'... *Organization* 10(2): 223–248.
Jones, C., M. Parker, and R. ten Bos. 2005. *For business ethics.* Oxon: Routledge.
Kant, I. 2008. *The critique of practical reason*, trans. J.K. Abbott. Forgotten Books. Available online at: www.forgottenbooks.org.
Keleman, M., and T. Peltonen. 2001. Ethics, morality and the subject: The contribution of Zygmunt Bauman and Michel Foucault to 'postmodern' business ethics. *Scandinavian Journal of Management* 17: 151–166.
Keynes, J.M. 1953. *The general theory of employment, interest and money.* New York: Harcourt Brace Jovanovich.
Kilduff, M., and A. Mehra. 1997. Postmodernism and organizational research. *Academy of Management Review* 22(2): 453–481.
Linstead, S. 2004. Introduction: Opening up paths to a passionate postmodernism. In *Organization theory and postmodern thought*, ed. S. Linstead, 1–13. London: Sage.
Lyotard, J-F. 1984. *The postmodern condition: A report on knowledge*, trans. G. Bennington and B. Massumi. Minneapolis: University of Minneapolis Press.
MacIntyre, A. 1984. Does applied ethics rest on a mistake? *The Monist* 67: 498–513.
Munro, R. 1998. Ethics and accounting: The dual technologies of the self. In *Ethics and organizations*, ed. M. Parker, 197–220. London: Sage.
Nietzsche, F. 1995. *On truth and falsity in their extramoral sense*, trans. M.A. Mugge. In *Philosophical writings,* ed. R. Grimm and C.M. Vedia, 87–99. New York: Continuum.
Painter-Morland, M. 2006. Redefining accountability as relational responsiveness. *Journal of Business Ethics* 66: 89–98.
Painter-Morland, M. 2008. *Business ethics as practice: Ethics as the everyday business of business.* Cambridge: Cambridge University Press.
Parker, M. 1993. Life after Jean-Francois. In *Postmodernism and organizations*, ed. J. Hassard and M. Parker, 204–212. London: Sage.
Parker, M. 1998a. Against ethics. In *Ethics and organizations*, ed. M. Parker, 282–296. London: Sage.
Parker, M. 1998b. Introduction: Ethics, the very idea? In *Ethics and organizations*, ed. M. Parker, 1–14. London: Sage.
Plessner, H. 1972. Die anthropologische Dimension der Geschichtlichkeit. In *Sozialer Wandel, Zivilisation und Fortschritt als Kategorien der soziologischen Theorie*, ed. H.P. Dreitzel, 160–168. Neuwied: Luchterhand.

Proust, M. 1934. *Remembrance of things past,* vol. 5 trans. C.K. Scott-Moncrieff. New York: Random House.

Rawls, J. 1971. *A theory of justice*. Cambridge, MA: Harvard University Press.

Rorty, R. 1979. *Philosophy and the mirror of nature*. Princeton: Princeton University Press.

de Saussure, F. 1960. *Course in general linguistics*, ed. C. Bally and A. Sechehaye and trans. W. Baskin. London: Peter Owen.

Seabright, M.A., and L.B. Kurke. 1997. Organisational ontology and the moral status of the corporation. *Business Ethics Quarterly* 7(4): 91–108.

Sim, S. 2001. Postmodernism and philosophy. In *The Routledge companion to postmodernism*, ed. S. Sim, 3–14. London: Taylor and Francis.

ten Bos, R. 1997. *Essai*: Business ethics and Bauman ethics. *Organization Studies* 18 June, 1997: 997–1014.

Trevino, L.K., and G.R. Weaver. 1994. Business ETHICS/BUSINESS ethics: One field or two? *Business Ethics Quarterly* 4(2): 113–128.

Verstraeten, J. (ed.). 2000. *Business ethics: Broadening the perspective*. Leuven: Peeters.

Wakefield, N. 1990. *Postmodernism*. London: Pluto.

Walton, C.C. 1993. Business ethics and postmodernism: A dangerous dalliance. *Business Ethics Quarterly* 3(3): 285–305.

Willmott, H. 1998. Towards a new ethic? The contributions of poststructuralism and post-humanism. In *Ethics and organizations*, ed. M. Parker, 76–121. London: Sage.

Woermann, M. and Hattingh, J.P. 2008. Guidelines for teaching business ethics: Comparing experiences from South African universities. Unpublished Report, Stellenbosch University.

Wood, D. 1990. *Philosophy at the limit*. London: Unwin Hyman.

Chapter 2
The Ethics of Complexity and the Complexity of Ethics

Abstract In this chapter, a general and critical approach to complexity is introduced. Within this approach, complex systems are viewed as irreducible. In other words, it is deemed impossible to uncover the laws of complex systems; and, since complex systems cannot be fully modelled, any engagement with complexity necessitates a critical engagement with the limits and status of our knowledge claims. This critical engagement is denoted by the 'ethics of complexity'. Furthermore, the ethics of complexity commits us to a complex view of ethics. This is because complexity is inherent to any ethical engagement, yet ethical frameworks are also models; and, like all models, are limited, exclusionary, and incapable of accounting for the complexity of lived phenomena. However, models are also necessary, since we need to reduce the complexity, in order to make sense of our world. It is therefore argued that the best ethical models are those that draw attention to their own limited status, and, in this vein, the provisional imperative – which is a self-undermining imperative – is introduced as a guide for responsible ethical action.

Introduction

The French philosopher and sociologist, Edgar Morin (2008:16) argues that the term 'complexity' acts as 'a warning to our understanding, a cautioning against clarification, simplification, and overly rapid reduction'. This warning – especially as applied to the discipline of business ethics – served both as the impetus for this study, and will again be reflected in the insights offered in the conclusion to the study. However, since 'complexity' is often used as a conceptual catch-all to describe things that lack simple explanations, a more nuanced understanding of the concept is necessary, in order to develop a position that is both meaningful and useful. More specifically, through a careful elucidation of the ideas that inform complexity thinking, I hope to show how the warning contained in this type of

M. Woermann, *On the (Im)Possibility of Business Ethics*, Issues in Business Ethics 37, DOI 10.1007/978-94-007-5131-6_2, © Springer Science+Business Media Dordrecht 2013

thinking serves as a challenge to the underlying assumptions that underscore existing conceptual paradigms, and thereby also challenges the whole system of thought that defines our views on practice, politics, and ethics.

Of specific interest, are the implications that complexity thinking holds for our understanding of ethics. Some of the implications introduced in the previous chapter can be framed in the language of complexity, and this re-inscription serves as a useful starting point for carrying the analysis forward. In particular, the collapse of the fact-value distinction holds illuminating insights for how we are to think about both the *ethics of complexity* and the *complexity of ethics*.

As will be argued in more detail further on in this chapter, from the perspective of a general understanding of complexity, it is impossible to know phenomena in their full complexity (as is commonly assumed by strong relativists, who study systems in terms of their component parts). This means that any engagement with complexity thinking implies a critical engagement with the *status* and *limits* of our knowledge claims. The implications of this are twofold. Firstly, if we cannot have complete knowledge of complex things, we cannot calculate their behaviour in any deterministic way. When describing the world, we should account for the status of our interpretations and evaluations; or, otherwise stated, for the way in which we theorise our descriptions. Our modelling choices are based on our contextual and contingent understanding of the world, and on subjective judgements about what matters – both in our work and in our private lives. Consequently, our models should be viewed in both a descriptive and normative light. Engaging with the *ethics of complexity* is thus a structural condition of complexity thinking.

Secondly, although complexity thinking necessarily involves ethics, normative considerations cannot be justified on *a priori* grounds, because all ethical positions are contingent, limited, and exclusionary; and reflect the biases of the theorist. Moreover, since complexity prevents us from developing and defending a universal ethical position, ethical analysis denotes a serious engagement with, and evaluation of, how we describe theory. In order to promote responsible thought and action, our accounts of ethics should be open to revision and be subject to a deconstruction of sorts. This is the only way in which we can pay homage to the *complexity of ethics*, to which the ethics of complexity commits us.

In order to better understand, and develop the implications of this argument, it is necessary to provide a more comprehensive overview of the type of complexity thinking that informs this argument.

Characterising Critical Complexity

Two Understandings of Complexity Theory

A very good overview of the development of, and different interpretations given to, complexity thinking is to be found in Morin's (2007) article entitled 'Restricted complexity, general complexity'. In this article, Morin juxtaposes complexity thinking with the epistemology of the classical sciences. This epistemology is governed by the explanatory principle of reduction, and is supported by both the principle of disjunction (which entails the separation of cognitive difficulties from one another) and the principle of universal determinism (which constitutes the Laplacian idea that deterministic processes govern past, present, and future cosmic events). This classical epistemology is challenged by a number of recent developments in the fields of physics, mathematics, biology, economy, engineering, and computer science – all of which have contributed to the emergence of complexity thinking. It is beyond the scope of this chapter to explore these developments, and the influential movements that serve as predecessors to complexity thinking (most notably, cybernetics, general systems theory, and chaos and catastrophe theory). However, one theoretical distinction, which highlights both the persistence of classical epistemology, as well as the different approaches to complexity, is the distinction between *restricted* and *general* complexity. The central difference between these approaches concerns our *attitude* towards modelling complexity.

Restricted Complexity

According to Morin (2007), the goal of restricted complexity is to study the multiple, interrelated processes that constitute complex systems, in order to retroactively uncover the rules or laws of complexity. This approach is popular amongst the researchers at the Sana Fe Institute (which was founded in 1984, and which is dedicated to the study of complex systems), and much of the work conducted at the Institute is dedicated to discovering, comprehending and communicating the *common fundamental principles* in a variety of systems, which underlie many of the pressing problems currently facing science and society.

Whilst research conducted at the Institute has no doubt led to important advances in formalisation and modelling, Jack Cowan (in Horgan 1995: 104), one of the Institute's founders, notes that the major discovery to have emerged from the Institute is the insight that 'it's very hard to do science on complex systems', if by science one understands the process of discovering and modelling the rules and laws that govern the behaviour of all phenomena. Such a view is informed by, what John Horgan (1995) calls, a seductive syllogism, which is based on the premises that since a computer that follows a simple set of mathematical rules can give rise to extremely complicated patterns, and since extremely complicated patterns exhibit in the world, simple rules must also give rise to complicated worldly phenomena.

In their implicit acceptance of this syllogism, many researchers at the Institute, thus adhere to a *restricted* approach to complexity. Morin (2007) argues that even though these researchers work with dynamical systems, constituted by a large number of interactions and feedback (which serve to frustrate efforts at prediction and control), they still remain locked in the epistemology of the classical sciences, because:

> When one searches for the "laws of complexity", one still attaches complexity as a kind of wagon behind the truth locomotive, that which produces laws... Actually, one avoids the problem of complexity which is epistemological, cognitive, paradigmatic. To some extent, one recognizes complexity, but by decomplexifying it (10).

General Complexity

In contrast to this restricted view, the *general* perspective of complexity constitutes an engagement with complexity itself (as well as with the epistemological implications that arise from such an engagement). Although we cannot conclusively state that complexity is an ontological category of the world, as opposed to merely a consequence of our epistemological limitations, this does not imply that we can merely relegate complexity to the status of a *practical* problem (that can ultimately be solved with enough computing power). One cannot simply 'cut-up' complex systems in order to understand them, since what is of interest is the dynamic, local interrelations that exist between the parts of a complex system, and that give rise to emergent phenomena (which are often not reducible to base laws). In this process, contingency – expressed in terms of both intra-systemic and extra-systemic conditions (Wimsatt 2008) – also plays a crucial role, which further frustrates any efforts to merely calculate the resultant effect of a certain configuration of parts.

In terms of ethics, those who believe that categorically-binding ethics (and hence a logical and lawful view of the moral world) is defensible, follow a restricted approach to ethics, whereas those who seek to develop a complex understanding of the moral world, and actively engage with this complexity, follow a general approach to complexity. Since the general view of complexity is compatible with the position being developed in this study, this view will be further explained and illustrated in the next section, before exploring the ethical implications that arise from this view. This view will also henceforth be denoted by the term 'complexity'.

Features of Complex Systems

The paradigm of complexity claims a certain universality, in that it deals with systems as such, and not just with component parts of systems. However, the paradigm of complexity – like all models – is limited, and therefore can 'neither... reflect the complete reality of its object, nor... exhaust all the possibilities of knowing its object' (Luhmann 1995: xlvii). The complexity and management theorist, Peter Allen (2000), describes this difficulty in terms of a paradox: on the one hand we wish to gather systemic knowledge about the objects under study; yet, on the other hand, these objects are impossible to unravel (due to the complexity). Moreover, complex systems are contextually-defined, and therefore any attempt at defining the general characteristics of complex systems should merely be an indication of the type of considerations one should keep in mind when studying these systems, rather than constitute a detailed description of a particular system. These *caveats* aside, we can now attempt a characterisation of the features of complex systems.

Complex Systems Are Not Complicated Systems

In avoiding the problem of complexity, the restricted approach is based on the implicit assumption that phenomena in the world are merely complicated, as opposed to truly complex. Kurt Richardson (2001, 2002) and Cilliers (1998) both provide a good overview of the differences between complicated and complex systems. Cilliers describes the central difference as follows: whereas a complicated system may initially look complex (due to the large number of components that constitute the system, and/or the sophistication of the tasks that the systems can perform), the hallmark of a complicated system is that it is – in principle – *solvable*. In other words, given enough information and resources, the dynamics of a complicated system can be fully understood.

As mentioned, the complex nature of certain systems (especially living systems) is attributable to the dynamic, self-organising relationships and feedback loops that exist between components, and that – over time – give rise to systemic structures. In other words, certain systems are inherently complex due to their *organising processes*. Since only certain aspects of complex systems can be understood at a given time (depending on how we model the system), it also means that complex systems are not compressible. In this regard, the theoretical biologist, Robert Rosen (1985: 424), argues that a system is complex precisely 'to the extent that it admits non-equivalent encodings; encodings which cannot be reduced to one another'. An example that has often been used by Cilliers to illustrate the difference between complicated and complex systems is that of a jumbo jet and mayonnaise: whereas the former is complicated, the latter – as all chefs can attest to – is complex!

Admittedly, the distinction between complicated and complex systems is often undermined in practice by powerful new technologies, where, on further inspection, complex phenomena turn out to be merely complicated (Cilliers 1998). However, despite the fact that the distinction between complicated and complex systems cannot be drawn in any unproblematic manner, the distinction nevertheless remains a useful analytical tool as it determines whether the study of complexity constitutes a search for underlying mathematical rules and formulae, or whether the study of complexity constitutes a considered engagement with complexity. Whereas complexity theorists following the restricted approach implicitly accept the scientific ideals of explanation, prediction, and facilitation of control (Chu et al. 2003), complexity theorists following the general approach try to develop strategies and models to help us better deal with the complexities that characterise not only living systems, but also social systems.

Complex Systems Are Characterised by Richly Interconnected Components

Complex systems consist of a large number of interrelated components. According to Cilliers (1998), the interactions between the components can be physical or informational; they are fairly rich, meaning that 'any element in the system influences and is influenced by quite a few other ones' (3); they have a short range, (however, these local interactions can have large systemic effects, which implies that systems-level order emerges because of interactions amongst components at lower levels of the system); and, there are positive (stimulating) and negative (inhibiting) feedback loops in the interactions.

In simple phenomena the system is the additive result of its components. However, in complex phenomena, the system is the result of these nonlinear, dynamic interrelations between component parts (Cilliers 1998). This insight has important methodological implications:

As argued above, classical science is based on a reductive methodology in that it is believed that systems can be completely understood in terms of their components parts. Such analyses however serve to destroy the complexity, since what is of interest in complex systems is not the components themselves, but the interrelations between component parts. In response to the problems caused by reductionism, many have turned to the system (rather than its constituting parts) as the object of analysis in the hope of unearthing those principles that are common to all systems (see Von Bertalanffy 1972). However, this approach is equally problematic, as it is based on the principle of holism, which, as Morin (1992: 372) explains, is also a form of reductionism:

> Holism is a partial, one-dimensional, and simplifying vision of the whole. It reduces all other systems-related ideas to the idea of a totality, whereas it should be a question of confluence. Holism thus arises from the paradigm of simplification (or reduction of the complex to a master-concept or master-category).

In order to understand complex systems, we therefore need to account for both the systemic identity of component parts, and the complex nature of interrelations between both the component parts and between the component parts and the system as a whole.

The Component Parts of Complex Systems Have a Double-Identity

The component parts of complex systems have a double-identity, which is premised on both a diversity and a unity principle (Morin 1992). With regard to the diversity principle, the identities of the system's components are irreducible to the whole, since each component still retains its own unique individual identity. For example, the fact that I enjoy painting cannot be deduced from my role as a philosophy lecturer at a university, although it is part of what makes me unique. However, the coupling of components also gives rise to a common identity (the unity principle) which constitutes their citizenship in a system. Therefore, the fact that I interact with other academics and students on a professional basis, constitutes behaviour that supports the goals of the university, and thus confirms my identity as a philosophy lecturer in the academic system. This point applies generally: when thinking about complex systems, this double-identity needs to be accounted for, because – on the one hand – if we forego the diversity-principle, our thinking becomes increasingly homogenised (holism); and – on the other hand – if we forego the unity-principle, our thinking loses a sense of unity (Morin 1992).

Upward and Downward Causation Give Rise to Complex Structures

Competitive and cooperative interactions between component parts on a local level, give rise to self-organisation, defined as 'a process whereby a system can develop a complex structure from fairly unstructured beginnings' (Cilliers 1998: 12). Consider, for example, a group of students who come together to form a reading group on complexity theory. Through cooperating, their actions give rise to certain structures, thereby transforming their uncoordinated individual activities into coordinated, goal-orientated group activities. In turn, these self-organising processes feed back to constrain the behaviour of the parts through a process of downward causation. In my example, the behaviour of the individuals in the reading group is constrained by the goals of the group itself. In other words, in the context of the reading group, it would be inappropriate to instigate a conversation on my vacation plans. This underscores the point that, in order to understand complex phenomena, we must substitute the principle of reductionism with a principle that conceives of whole-part mutual interaction (Morin 1992).

These mutual interactions result in, what Morin (2008: 49) terms, 'organizational recursion' where 'the products and the effects are at the same time causes and producers of what produces them.' This means, for example, that, as individuals, we create, engage in, and challenge our practices (including our business practices), which simultaneously serve to shape us, as individuals.

Complex Systems Are Non-additive

The complex interrelations between components and systems, as described above, give rise to the following three systemic characteristics (Morin 1992): Firstly, the whole is greater than the sum of its parts, in that systemic attributes cannot be reduced to the parts alone, but are the result of interconnections between the parts. This is a widely recognised implication of systems analysis. Secondly, and less widely recognised, is the fact that the whole is also less than the sum of its parts, since some of the qualities of the parts are suppressed under the constraints that result from systemic organisation. The example previously mentioned pertains: the fact that I like painting is not an attribute that I can develop in my work, given my function as a philosophy lecturer at the university, yet this like also defines me, and (along with many other personal attributes) contributes to a personal identity that supersedes any given role identity. Thirdly, and perhaps most counter-intuitively, the whole is also greater than the whole, due to the dynamic organisation that takes place is systems where local interactions between components give rise to phenomena that are dependent on the base, but simultaneously supersede that base. In other words, the whole is greater than the whole because of self-organisation and systemic emergence.

Complex Systems Exhibit Self-Organising and Emergent Behaviour

In order to make the case for self-organisation, it is necessary to show that 'internal structure can evolve without the intervention of an external designer or the presence of some centralised form of internal control' (Cilliers 1998: 89). Self-organisation is a necessary condition for emergence, which is defined as 'the idea that there are properties at a certain level of organization which cannot be predicted from the properties found at lower levels' (Emmeche et al. 1997: 83). Specifically, self-organisation draws attention to the structural and temporal dimensions of emergence. Systems develop structure (i.e. hierarchies) by processing information and developing 'memory'. The example of neural networks offers a good explanation of this principle: neural networks are chemically-connected or functionally-associated neurons. The interconnections between these neurons are called synapses. Over time, certain pathways are established in the brain, meaning that some of the synapses are reinforced through impulses, whereas others die off. In this way, structure develops as groups of neurons are selected, reinforced, and transformed through interaction with their environment. This implies that a fairly undifferentiated brain develops structure or consciousness over time (Cilliers 1998).

This example adequately illustrates why, in complex systems, the whole is greater than the whole, as one can convincingly argue that the mind is 'greater' or 'more' than the brain (which is made up of self-organised neurons or synapses). It therefore seems that, whilst self-organisation is a necessary condition for emergence, it is not sufficient. However, despite this, there is still much debate regarding the nature of emergence (in this regard, see Bedau and Humphreys' (2008) jointly edited book, entitled 'Emergence: Contemporary Readings in Philosophy and Science').

Complex Systems Are Structured

When the components of systems interact, dynamic structures emerge over time due to self-organisation. The analysis thus far serves to dispel the popular notion that complex systems are flat systems. Only homogenous or chaotic systems are flat, because the complex processes that lead to the emergence of dynamic structures are absent (Cilliers 2001). Complex systems also contain a number of nested systems. However, one should remain cognisant of the fact that a nested system is very much the product of the description that one gives to the system. For example, a department can simultaneously be viewed as a nested system within a larger organanisational system, or it can be defined as a system in its own right, depending on one's level of analysis or framing strategy. Nonetheless, structures are indispensible for systemic development, since, as Cilliers (1998: 89) argues, the structure constitutes 'the internal mechanism developed by the system to receive, encode, transform and store information on the one hand, and to react to such information by some form of output on the other.' The nature of systemic structures are however very much contingent on the particularities of the system under study: whereas some structures are more durable (for example, bureaucratic structures within organisations), others are more volatile and ephemeral (for example, the structural relations between members of a working group) (Cilliers 2001).

Complex Systems Are Open and Bounded Systems

Not only are the particularities of the system under study vital for understanding the intelligibility of complex systems, but so too is the system's relation with the environment. This is because there is an energy, material, or information transfer into or out of a given system's boundary. The nature and content of the system's interaction with the environment is discipline-dependent. Whereas simple, homeostatic systems (such as a thermostat) are merely dependent on the environment for their survival in that they are capable of facilitating their own production and maintenance through feedback loops (Maturana and Varela 1980), human identity, for example, is in part constituted by the environment. This is because our identities develop over time within a network of relationships with other identities. As such, who I become (i.e. my emergent identity) is not only a function of my genes, but also of my context.

Regardless of the system under study, we can say that, methodologically-speaking, it is very difficult to study open systems. This is because the environment is simultaneously intimate and foreign: it is both part of the system (in that we reproduce the system-environment distinction when we model) and remains exterior to the system (Morin 2008). In other words, the environment cannot be appropriated by the system. This means that the boundary between a system and its environment should be treated both as a real, physical category, and a mental category or ideal model (Morin 1992). This last point has implications for how we view the boundaries of systems: although boundaries are a function of the activity of a system itself, they are also the product of the description that we give to the system. Hence, boundaries should be thought of 'as something that constitutes that which is bounded' (Cilliers 2001: 141), rather than as an objective demarcation of a system.

The fact that we cannot draw a system's boundary in any unproblematic fashion introduces further complexities, as denoted by the observer problem. Niklas Luhmann (2000: 46) describes this consequence of modelling (or bounding systems) as a paradox:

> The self-description of the self-transparent system has to use the form of a paradox, a form with infinite burdens of information and it has to look for one or more distinguishable identities that "unfold" the paradox, reduce the amount of needed information, construct redundancies, and transform unconditioned into conditioned knowledge... [but] the question of the unity of the distinction always leads back to the paradox – and one can show this to others and accept it for oneself.

The issue at stake is not so much the paradox itself: if we concede to the fact that the world is complex, then the paradox of framing or modelling is part of the complexity with which we have to grapple. In other words, we have to frame. Rather, the issue is whether we accept and account for the paradoxical status of frames in our theories, which amounts to the fact that we draw boundaries in order to lessen the complexity, but in so doing, we add another level of distortion and complexity. This paradox is especially pertinent in the human sciences, where any model of human behaviour immediately feeds back into the system and alters the behaviour of the subjects under study – a phenomenon denoted by the term, 'the double hermeneutic loop'.

Ethical Implications

The Ethics of Complexity and the Limits of Knowledge

The Status of Our Models

In the introduction to this chapter it was stated that engaging with complexity challenges current conceptual paradigms, and thereby necessitates an epistemological shift. This means that we should rethink the nature, status, and limits of our knowledge claims. Grappling with complexity cannot constitute a purely objective exercise; however, just like an affirmative account of postmodernism does not commit one to a view that 'anything and everything goes', so too, productively engaging with complexity also constitutes an attempt to build systemic knowledge. As Peter Allen (2000: 93) states, '[a] representation or model with no assumptions whatsoever is clearly simply subjective reality'. For him, working with the irreducible nature of complexity means that we should 'apply our "complexity reduction" assumptions honestly' (94), rather than accept the defeatist attitude that limited knowledge commits us to relativism. Again, this brand of intellectual honesty implies a modest attitude, and denotes sensitivity to the levels and limits of our understanding. We should still perform the necessary calculations and make the necessary reductions, but we should recognise that such activities can lead to the development of useful models, not resolve the complexity. This is because each modelling choice gives rise to 'a different spectrum of possible consequences, different successes and failures, and different strengths and weaknesses' (102). Knowledge acquisition should not be viewed as the objective pursuit of truth, but rather as the attempt to develop models that accord with our experiences in the world. In this regard, a model is more like a novel than a formula, in that:

> like a novel, [a model] may resonate with nature, but it is not the "real" thing. Like a novel, a model may be convincing – it may "ring true" if it is consistent with our experiences of the natural [and the social] world. But just as we may wonder how much the characters in

> a novel are drawn from real life, and how much is artifice, we might ask the same of a model: how much is based on observation and measurement of accessible phenomena, how much is based on informed judgement, and how much is based on convenience? (Oreskes et al. 1994: 644).

Models, like boundaries, do not merely create distortions: they are also enabling, because they help us to make sense of our world. However, since there is no final model, we should introduce a double consciousness into our models: 'a consciousness of [models] and an ethical consciousness' (Morin 2007: 21). This double consciousness is necessary precisely because our models are contingent and limited, but materially affect our practices.

Reductionism in the Social Sciences

In his article, entitled 'Bad management theories are destroying good management practices', Sumantra Ghoshal (2005: 75) affirms this point in arguing that 'we – as business school faculty – need to own up to our own role in creating Enrons. Our theories and ideas have done much to strengthen the management practices that we are all now so loudly condemning.' Ghoshal specifically argues against the causal and functional nature of management models, arguing that this approach has led both to the pretence of knowledge, and to the denial of ethical and moral considerations in business theories (as such considerations do not sit well with the pseudo-scientific approach that underscore most management models). As a result of this strong reductionist tendency in management theory, managers actively free themselves from assuming moral responsibility for their actions. Ghoshal illustrates the effects of this pretence of knowledge as follows:

> when managers, including CEOs, justify their actions by pleading powerlessness in the face of external forces, it is to the dehumanization of practice that they resort. When they claim that competition or capital markets are relentless in their demands, and that individual companies and managers gave no scope for choices, it is on the strength of the false premise of determinism that they free themselves from any sense of moral or ethical responsibility for their actions (79).

Allen (2000) also discusses the problem of strong reductionism in the management sciences, and identifies the following five reductionist assumptions: (1) we can clearly define the boundaries between the system and the environment; (2) we already possess rules needed to classify objects into relevant typologies, which will enable us to understand what is going on; (3) individuals are considered as average types that are not affected by experiences; (4) individual behaviours can be described by their average interaction parameters; and, (5) stability or equilibrium defines reality. Many of these same reductions are made in business ethics, where it is assumed that the rational individual is capable of engaging in abstract ethical thought, and thereby capable of coming to appropriate ethical conclusions that can then be applied to a situation.

A problem with this approach to the social sciences is that (in appropriating the scientific paradigm), social science is often construed as a value-free programme, aimed at establishing *a priori* laws and rules through reductive reasoning. This is because such approaches not only negate complexity, but also the double consciousness needed for responsibly engaging in ideas and modelling experience.

Modelling and the Importance of a Double-Consciousness

This double consciousness, to which Morin (2007) refers, should not only inform our development and assessment of models, but also our assessment of the practices that stem from these models. Although we should exercise vigilance when choosing our strategies, we should also recognise that no matter how carefully we consider and reflect upon these actions, they may turn out to be a mistake (Preiser and Cilliers 2010). Morin (2008: 55) argues that if our actions 'fly back at our heads like a boomerang', we are obliged to follow the actions, in order to attempt to correct them. This will only be successfully achieved if we are aware from the outset that action implies 'risk, hazard, initiative, decision... derailments and transformation.' He also states that every 'action is also a wager' and '[w]e must be aware of our philosophical and political wagers' (54). When we refuse to engage in the complexity with which we are confronted when we undertake decisions and actions, and when we close down awareness of what lies outside of our immediate experiences, we are in danger of irresponsible action.

Recognising the *ethics of complexity* is a means of expressing this double-consciousness, because, firstly, the ethics of complexity compels us to accept and account for the status of our models i.e. the fact that models are largely an outcome of judgement and convenience (especially convenience in terms of utilising the resources at our disposal for creating these models). In other words, acknowledging the ethics of complexity leads to conscious, and moreover self-conscious modelling. Secondly, in acknowledging the limited status of our models, and the ways in which these imperfect models materially affect our lives and the lives of others, the ethics of complexity compels us to remain perpetually vigilant, in order to promote responsible thought and action. Vigilance demands a continual and critical evaluation and transformation of our claims and practices, and commits us to a radical or recursive view of responsibility, defined as an 'always renewable openness' (Wood 1999: 117). This renewable openness is safe-guarded by a self-critical rationality, which Cilliers (2010a: 14) defines as 'a rationality that makes no claim for objectivity, or for any special status for the grounds from which the claim is made', and which he describes as the *outcome* of acknowledging the irreducible nature of complexity. The ethics of complexity therefore not only commits us to a general, but also a *critical* understanding of complexity.

The Embeddedness of Ethical Practices: Positioning the Moral Agent

Above, it is argued that a complexity-based understanding of the world commits us to accepting responsibility for the outcomes of our models, especially with regard to how these models affect our lived practices. However, a second consequence of this complexity-based understanding concerns the implications that such a view holds for how we are to think about ontology or the identity conditions of phenomena in the world, including the identity conditions of the moral agent.

Static Versus Fluid Conceptions of Identity

As previously stated, much of the (business) ethics literature ascribes moral responsibility to rational and autonomous individual agents who make decisions based on reasonable principles and calculations. It is also commonly assumed that these moral agents act intentionally. In other words, it is assumed that there is 'a direct cause and effect relationship between the willing and acting agent and the consequences of his or her decisions and behavior' (Painter-Morland 2006: 90). In this view, individuals are identified as ontologically *prior* to the systems in which they function, due to the fact that the identity conditions of individuals (namely, intentionality, autonomy, and rationality) are assumed to be *a priori* givens. What should be clear at this juncture however is that individual identity is a relational and emergent property, and should be contextualised in terms of a spatial network of relationships in which individuals are co-constituted, as well as in terms of a temporal process of becoming (Woermann 2010; Cilliers 2010b).

Following the complexity insights introduced earlier, this implies that the focus of any ethical analysis should not be exclusively on individual agents, but also on the *relations* between individuals and the systemic properties that emerge from these relations. Through participating in competitive and cooperative activities, the intentional and unintentional actions of individuals give rise to certain patterns of behaviour, which in turn lead to the emergence of systemic structures (in this regard, consider again the example of initiating a complexity reading group). Systemic structures, for their part, serve to constrain the behaviour of individual components through feedback loops, but also create opportunities in the structure, and thereby facilitate purposive action. For example, belonging to an organisation constrains the type of activities that are deemed appropriate to undertake during the course of one's work life, but also affords one opportunities to develop one's business acumen and to realise projects that cannot be undertaken in an individual capacity, due to limited resources and/or expertise.

Over time, these feedback loops reinforce or undermine certain patterns of behaviour, which then become institutionalised in formal or informal rules, norms, policies, laws etc. In organisations, formal instruments such as mission and vision statements, codes of conduct and organisational policies, as well as informal norms – as expressed in an organisation's cultural artifacts – serve the purpose of aligning individual and organisational goals and expectations.

However, due to the complexity at play, radical systemic transformations are also possible. Consider, for example, former group executive of BP, Tony Hayward's decision to go sailing during the recent oil spill in the Gulf of Mexico, which contributed to his perceived lack of leadership during the first months of the catastrophe, and which ultimately led to his dismissal. Here we see how one unforeseen event (i.e. the oil spill) served to change not only Hayward's destiny, but also impacted on the leadership structures and priorities of BP and the oil industry. In this regard, Bob Dudley, the new group executive, stated on the company website that: 'We have learned – and will continue to learn – many lessons from this oil spill. The many investigations of the accident will bring changes to our industry – changes that will improve the safety of deepwater drilling going forward.'

Moral Agency in a Complex World

The above analysis reaffirms the point made in the previous chapter that the questions 'what ought I to do?' and 'what type of person should I be?' are inseparable, given the embeddedness of our practices and identities. In other words, individuals' role identities and the systems within which they operate are *co-terminus* i.e. they arise and die together (see Seabright and Kurke 1997). How we act within these systems is therefore also contingent upon the roles that we have within these systems, and how these systems serve to enable and constrain what is deemed appropriate and inappropriate behaviour. We also however possess an identity that supersedes any given role identity (in complexity terms, the whole is also less than the sum of its parts). If this were not possible, intentionality would only amount to living out our roles, as defined within a given system. On a complexity view, the space of free will is not vacated; however, what is discredited is the portrayal of the moral agent 'as an independent or socially isolated decision-maker' (Kramer 1991: 195). The moral agent should rather be understood 'as a social actor embedded in a complex network of intra- and inter-group relationships' (195). This means that if we wish to talk about intentionality and moral agency, we can only do so by considering the specific 'context of relevant group memberships, the systems within which groups are embedded, the power relations that exist between groups and the permeability of boundaries that define group membership' (Paulsen 2003:17).

Although there cannot be a direct cause-and-effect relationship between the actions of an intentional agent and the consequences of her actions upon a system, we can nevertheless help shape and transform our systems through critical engagement expressed in communicative processes, in which the relative power and political skill of the various actors are considered. This is a very worthwhile process, specifically because we help to constitute our systems and one another through our interactions. In this vein, Preiser and Cilliers (2010: 268) argue that '"[e]thics" is not something that gets integrated into organisational or corporate culture, but lies at the heart of establishing a culture to begin with; it is part of all the different levels of activities in an organisation'. Furthermore, because of the complexity, it is also possible for small causes to have large effects (although it is impossible to predict this beforehand), and therefore we have 'to tinker often, and insistently' (Painter-Morland 2008: xi).

The Complexity of Ethics and Responsible Action

Towards a Meta-Ethical Position

Two important insights that have emerged from the discussion on ethics and complexity are that we should be cognisant of the limited status of our models, which is due to our incomplete understanding of complex phenomena; and, the moral agent (and hence responsible action) should be understood in terms of dynamic, emergent, and relational practices and group dynamics. Both these insights imply that, since no objective account of reality exists, all knowledge claims should be accepted as *provisional* claims, because, in the words of Preiser and Cilliers (2010: 270): 'We *know* that we cannot get it right.'

This insight is also acknowledged in postmodernist accounts; and, as argued in Chap. 1, gives our theories a distinctive anti-foundational slant. This is because acknowledging the ethics of complexity cannot amount to more than an *awareness* of the fact that we are always in trouble. In other words, acknowledging complexity precludes the formulation of substantive guidelines for an ethical system, precisely because there is no meta-position that we can appeal to, in order to help us out of this trouble. As Derrida (1988: 119) muses, '[i]f things were simple, word would have gotten around'.

Yet, the question that nevertheless arises is whether it is possible to move the analysis beyond this point, in order to say something more about the *complexity of ethics*. Preiser and Cilliers (2010) argue that, despite not being able to develop a substantive ethics, we can nevertheless develop a type of meta-ethical position, which can serve to highlight important ethical considerations that underscore the ethical strategies that we employ when engaging in the particularities of situations. For this task, they draw on perhaps the most famous example of a meta-ethical position in the history of moral philosophy, namely Kant's (1993) categorical imperative. This meta-ethical position is further unpacked in Woermann and Cilliers (2012: 450–452), and the analysis given in this latter paper is replicated below:

The Provisional Imperative

The categorical imperative is a substantively-empty rule, in that it cannot generate contingent ethical principles, but can merely act as a yardstick for evaluating the morality of principles that already exist. This is because Kant wants his moral rule to be categorically applicable and, hence, universally valid. However, the only rule which conforms to this criterion is a purely abstract and formal rule, which says 'always follow only universal rules'; or, otherwise stated, 'always follow only rules which you will want all other people to follow'. Thus, by combining a purely formal rule with the notion of universability, Kant can generate a formulation that actually does say something about ethics, namely that if certain contingent principles are universalisable, then the principles are deemed morally correct. Therefore, although the categorical imperative cannot indicate what principles are good, right, and deserving of respect, it does provide a strategy for evaluating our contingent principles. As such, one can argue that Kant's categorical imperative urges us to adopt a certain strategy when undertaking moral considerations (Preiser and Cilliers 2010).

Next, we can try to apply the same logic Kant uses to the ethics of complexity, in order to say something about the complexity of ethics (in other words, in order to develop our meta-ethical position). From the analysis thus far we can construct the following argument: all knowledge (including self-knowledge) is limited because, in order to generate meaning, we need to reduce the complexity through modelling. Our models are radically contingent in time and space because they are the product of the resources at our disposal, the choices that we make, and the influences that act upon us (including the influences of others). Since all knowledge is contingent, it is also subject to revision, and therefore irreducibly provisional. Following Kantian logic, we can now capture the gist of the above argument in the following imperative: 'When acting, always remain cognisant of other ways of acting'. The espoused meta-ethical position thus constitutes a *provisional imperative* (Preiser and Cilliers 2010).

Note that on one reading, the idea of a provisional imperative is a contradiction in terms, since the logic of an imperative is absolute: either you follow the imperative or you don't. The idea of a provisional imperative seems to suggest that the imperative itself is subject to change, and in this regard we seem to be advocating an impossible position. This is, to a large extent, exactly the point: we cannot do away with moral imperatives, but complexity thinking highlights the fact that our imperatives are the outcome of certain framing strategies or ways of thinking about the world, and are thus necessarily exclusionary. Thus the provisional imperative stipulates that we must be guided by the imperative, whilst simultaneously acknowledging the exclusionary nature of all imperatives.

In terms of the actual content of the imperative, it should be noted that – unlike the Kantian imperative – which tells us something about the rules for action, the provisional imperative says something about our state of mind or *attitude* when choosing rules for action. Again: it is impossible to say that 'When acting, always choose rules that admit to the possibility of other rules', since the logic of rules (as with the logic of imperatives) is absolute. In this regard, Derrida (1988: 116) notes that:

> Every concept that lays claim to any rigor whatsoever implies the alternative of "all or nothing"... Even the concept of "difference to degree," the concept of relativity is, qua concept, determined according to the logic of all or nothing, of yes or no: differences of degree or no differences of degree. It is impossible or illegitimate to form a philosophical concept outside the logic of all or nothing.

In the above citation, Derrida is pointing to a structural condition of all concepts. We cannot do other than model and exclude. Yet, what the provisional imperative shows is that, when we act, we should be cognisant of this logic. It makes a difference – and moreover a moral difference – whether one exercises this awareness. This is because if we remain open to other ways of modelling and other ways of being, then we are more likely to practice a self-critical rationality, to respect diversity, to be willing to revise our models, and to guard against the naturalisation of these models. These tasks are especially significant, given the nature of today's geopolitics in which Western ideals all-too-often pass as universal ideals. Here, it is useful to recall Derrida's (2002: 10) view of philosophy as something which is 'no more assigned to its origin or by its origin, than it is simply, spontaneously, or abstractly cosmopolitical or universal.' He continues in arguing that '[t]here are other ways for philosophy than those of appropriation or expropriation... Not only are there other ways for philosophy, but philosophy, if there is any such thing, is the other way'. In our context, we can substitute the term 'philosophy' with 'ethics', since what lies at the heart of the provisional imperative is the belief that ethics is indeed the other way; or, more poignantly, the way which is still to come. The provisional imperative therefore provides us with a strategy for remaining open to complexity at the same time that we reduce complexity through our decisions and actions.

Postmodernism, Complexity, and Theories of the Organisation

Although developed along different disciplinary lines, the paradigms of critical complexity and affirmative postmodernism hold similar consequences for our understanding of agency and of our social, intellectual, and ethical practices. As such, these paradigms reinforce and supplement each other. In terms of ethics, affirmative postmodernism and critical complexity constitute theoretical positions that are sensitive to the complex contingencies that impact on our idea of the good, and that consequently denounce self-sustaining or naturalised notions of the good.

A central insight that has emerged from the analysis thus far is that ethical problems are not limited to considerations of first-order normative principles. The account of affirmative postmodernism given in the first chapter shows that the rigid distinction between the normative and descriptive categories – or between what ought to be the case and what is the case – is untenable. The imbrication of the normative and the descriptive categories poses a huge challenge to our understanding of ethics, since we can no longer appeal either to a transcendental ideal realm or to an objective reality, when justifying our ethical positions. In other words – and as previously stated – the imbrication of ethics with the world introduces complexity.

In this regard, the analysis of critical complexity given in this chapter offers important insights for dealing with this postmodern re-description of ethics. From a complexity perspective, our identities are viewed as emergent and relationally-constituted constructs, rather than as *a priori* givens. The question of what it means to be human cannot therefore be divorced from an understanding of our immanent and embedded positions in the world. The paradigm of critical complexity can further help us to understand ourselves as component parts in a system, and for envisioning the potential consequences that our actions hold for other components in the system, and for the system as such. In terms of ethics, critical complexity provides the backdrop against which we can think through the status of our ethical models and the consequences that our decisions and actions hold; and, as argued, complexity thinking commits us to continuous and critical reflection on our practices and strategies.

Given the above argument, it follows that a complexity-informed understanding of the organisation differs from more traditional contributions in that the focus of analysis shifts from individuals (whether construed as managers, leaders, or ethical agents) to the processes and systems in which individuals partake. A number of theorists working in the fields of organisation and management studies, leadership, and business ethics have appropriated this basic complexity insight in their work; and, in this regard, it is worthwhile to take note of the following contributions: In the field of organisation and management studies, the work of Allen (2001) (which is focused on evolutionary complex systems and the limits to modelling knowledge) is significant, as is Michael Lissack and Hugo Letiche's analysis on the importance of coherence, emergence, and resilience for a complexity-based theory of organisations and organisational identity, and Ralph Stacey's work on complex responsive processes in organisations and management practices. In leadership studies, the work of Mary Uhl-Bien, Russ Marion, Bill McKelvey, Benyamin Lichtenstein and Donde Plowman should be noted. These theorists use complexity theory in order to develop a relational (as opposed to agent-centric) understanding of leadership, and to investigate the role of leadership in emergent, self-organising systems. In business ethics, Jane Collier and Rafael Esteban have made important contributions in terms of illustrating the role that complex processes play in ethical leadership and governance practices; Thomas Maak and Nicola Pless have explored the concept of a complex, relational form of leadership, and what this implies for ethical and responsive leadership practices; and, Painter-Morland has incorporated elements of complexity theory in her work on relational responsiveness, leadership, accountability, and business ethics management.

In focusing on the enabling and constraining effects that organisational systems and processes exert on individual behaviour, and how individuals in turn can affect organisational systems, these theorists seek to develop more complex models of organisational life. More specifically, these theorists have foregone a simple and predictive view of organisational life (which is necessarily 'bought' at the price of a number of reductive implications (Allen 2000), as discussed earlier), in favour of a theoretical approach that does not offer solutions in terms of what Preiser and Cilliers (2010: 289) term 'a "best practice" manual or toolkit'. Rather, they provide us with 'an integrated, multi-dimensional approach', which can lead to insights that can 'be related successfully to the real-world situation' (Allen 2000: 29).

Of specific interest in this study, is how a complexity-based understanding of the moral world can help us to better navigate through the complexities and challenges that define our working lives. Although a number of insights have already been explored in this chapter, the theoretical grounding for this analysis will be incomplete without an investigation of how these insights relate to our philosophical understanding of ethics. In terms of a philosophical ethics, it is proposed that Derrida's deconstructive ethics offers us a productive reading of a complex notion of ethics. This is because Derrida's position extends the postmodern insights introduced in Chap. 1, resonates well with the specific complexity insights developed in this chapter, and overcomes a number of challenges associated with pulling these complexity insights into the human domain – and more specifically, the domain of ethical decision-making. As such, a deconstructive ethics will be introduced in the next chapter.

References

Allen, P. 2000. Knowledge, ignorance, and learning. *Emergence* 2(4): 78–103.

Allen, P. 2001. What is complexity science? Knowledge of the limits to knowledge. *Emergence* 3(1): 24–42.

Bedau, M.A., and P. Humphreys (ed.). 2008. *Emergence: Contemporary readings in philosophy and science*. Cambridge, MA: The MIT Press.

Chu, D., R. Strand, and R. Fjelland. 2003. Theories of complexity: Common denominators of complex systems. *Wiley Periodicals* 8(3): 19–30.

Cilliers, P. 1998. *Complexity and postmodernism: Understanding complex systems*. London: Routledge.

Cilliers, P. 2001. Boundaries, hierarchies and networks in complex systems. *International Journal of Innovation Management* 5(2): 135–147.

Cilliers, P. 2010a. A provisional ethics of complexity. Unpublished.

Cilliers, P. 2010b. Difference, identity, and complexity. *Philosophy Today*, Spring 2010, 55–65

Derrida, J. 1988. Afterword. In *Limited Inc.*, ed. G. Graff and trans. S. Weber, 111–160. Evanston: Northern Western University Press.

Derrida, J. 2002. On the "priceless," or the "going rate" of the transaction. In *Negotiations: Interventions and interviews, 1971–2001*, ed. and trans. E. Rottenberg, 315–328. Stanford: Stanford University Press.

Emmeche, C., S. Koppe, and F. Stjernfelt. 1997. Explaining emergence: Towards an ontology of levels. *Journal for General Philosophy of Science/Zeitschrift für allgemeine Wissenschaftstheorie* 28(1): 83–119.

Ghoshal, S. 2005. Bad management theories are destroying good management practices. *Academy of Management Learning and education* 4(1): 75–91.

Horgan, J. 1995. From complexity to perplexity. *Scientific America* 1995: 104–109.

Kant, I. 1993. *Grounding for the metaphysics of morals*, 3rd ed., trans. J.W. Ellington. Indianapolis: Hackett.

Kramer, R.M. 1991. Intergroup relations and organizational dilemmas: The role of categorization processes. *Research in Organizational Behaviour* 13: 191–228.

Luhmann, N. 1995. *Social systems*, trans. J. Bednarz Jr. (with D. Baecker). Stanford: Stanford University Press.

Luhmann, N. 2000. Why does society describe itself as postmodern? In *Observing complexity: systems theory and postmodernity*, ed. W. Rasch and C. Wolfe, 35–49. Minneapolis: University of Minnesota Press.
Maturana, H.R., and F.J. Varela. 1980. *Autopoiesis and cognition: The realization of the living*. Dordrecht: D. Reidel.
Morin, E. 1992. From the concept of a system to the paradigm of complexity, trans. S. Kelly, *Journal of Social and Evolutionary Systems* 15(4): 371–385.
Morin, E. 2007. Restricted complexity, general complexity, trans. C. Gershenson. In *Worldviews, science and us: Philosophy and complexity,* ed. C. Gershenson, D. Aerts, and B. Edmonds, 5–29. Singapore: World Scientific.
Morin, E. 2008. *On complexity*, trans. S.M. Kelly. Cresskill: Hampton Press.
Oreskes, N., K. Shrader-Frechette, and K. Belitz. 1994. Verification, validation, and confirmation of numerical models in the earth sciences. *Science* 263(5): 641–646.
Painter-Morland, M. 2006. Redefining accountability as relational responsiveness. *Journal of Business Ethics* 66: 89–98.
Painter-Morland, M. 2008. *Business ethics as practice: Ethics as the everyday business of business*. Cambridge: Cambridge University Press.
Paulsen, N. 2003. Who are we now? Group identity, boundaries, and the (re)organizing process. In *Managing boundaries in organizations: Multiple perspectives*, ed. N. Paulsen and T. Hernes, 15–34. Hampshire: Palgrave Macmillan.
Preiser, R., and P. Cilliers. 2010. Unpacking the ethics of complexity: Concluding reflections. In *Complexity, difference and identity*, ed. P. Cilliers and R. Preiser, 265–287. Dordrecht: Springer.
Richardson, K.A. 2001. On the status of national boundaries: A complex systems perspective. *Proceedings of the Systems in Management 7th Annual ANZSYS Conference*, Edith Cowan University, Churchlands, 27–28 Nov 2001, 229–238.
Richardson, K.A. 2002. Methodological implications of complex systems approaches to sociality: Some further remarks. *Journal of Artificial Societies and Social Simulation* 5(2). Available online at: http://jasss.soc.surrey.ac.uk/5/2/6.html.
Rosen, R. 1985. *Anticipatory systems: Philosophical, mathematical and methodological foundations*. Oxford: Pergamon Press.
Seabright, M.A., and L.B. Kurke. 1997. Organisational ontology and the moral status of the corporation. *Business Ethics Quarterly* 7(4): 91–108.
Von Bertalanffy, L. 1972. The history and status of general systems theory. *The Academy of Management Journal* 15(14): 407–426.
Wimsatt, W.C. 2008. Aggregativity: Reductive heuristics for finding emergence. In *Emergence: Contemporary readings in philosophy and science*, ed. M.A. Bedau and P. Humphreys, 99–110. Cambridge, MA: The MIT Press.
Woermann, M. 2010. Corporate identity, responsibility and the ethics of complexity. In *Complexity, difference and identity*, ed. P. Cilliers and R. Preiser, 167–192. Dordrecht: Springer.
Woermann, M. and Cilliers, P. 2012. The ethics of complexity and the complexity of ethics. *South African Journal of Philosophy*, 31(2): 447–463.
Wood, D. 1999. The experience of the ethical. In *Questioning ethics: Contemporary debates in philosophy*, ed. R. Kearney and M. Dooley, 105–119. London/New York: Routledge.

Chapter 3
Introducing a Deconstructive Ethics

Abstract In this chapter, deconstruction is explained and the ethics of deconstruction is explored. It is specifically argued that Derrida's philosophy offers a productive reading of a complex notion of ethics, for the following three reasons: Firstly, Derrida's work on quasi-transcendental or limit concepts provides a means for addressing the methodological complexity of thinking together a system and its environment, and thereby both constitutes as engagement with the ethics of complexity, and serves to articulate the ethical interruption of ontological closure. Secondly, in deconstructing the conceptual models that inform our practices, Derrida is able to de-naturalise these models and thereby draw attention to both the ethics of complexity and the supplementary complications that pervade all meaning. In so doing, Derrida opens the door to otherness and difference, despite his radically immanent and contextualised understanding of ethics. Thirdly, in explicitly addressing the ethical-political implications that arise due to the limitations of our knowledge claims, Derrida's *oeuvre* allows for a sophisticated and thorough analysis of the implications that critical complexity thinking hold in the human domain. Derrida's work therefore lends philosophical grounding to the insights gleaned from the analysis on critical complexity thinking undertaken in the previous chapter.

Introduction

The term 'deconstruction' is notoriously difficult to define. Indeed, Derrida writes that '[a]ll sentences of the type "deconstruction is X" or "deconstruction is not X" *a priori* miss the point' (Derrida 1988a: 4). The problem with defining the term stems from the fact that all the predicates that lend themselves to the definition are also deconstructable, including the unity of the word itself: *'It deconstructs it-self'* (4). As soon as we have meaning, it can be deconstructed. In other words, deconstruction takes place whenever there is something (Critchley 1999a); and, conversely, if something is not deconstructable, then it is transcendental and, hence, impossible.

M. Woermann, *On the (Im)Possibility of Business Ethics*, Issues in Business Ethics 37, DOI 10.1007/978-94-007-5131-6_3, © Springer Science+Business Media Dordrecht 2013

The function of deconstruction is to seriously engage with, and challenge, the conceptual schemas that inform our understanding, in order to destabilise them. As will be explained in more detail in this chapter, this is done by revealing the metaphysical idealisations that are smuggled into our schemas, and by de-naturalising these positions through highlighting the supplementary complications that pervade all meaning. In so doing, Derrida opens the door to otherness and difference, and thereby shifts the limits of our understanding. In this regard, Derrida (1988a: 4) writes that deconstruction is 'the delimiting of ontology'. Simon Critchley (1999a: 30) further argues that deconstruction provides us with a *clôtural* reading, which 'articulates the ethical interruption of ontological closure, thereby disrupting the text's claims to comprehensive unity and self-understanding'.

This brief description of deconstruction already alludes to the fact that deconstruction represents a serious philosophical engagement with both the ethics of complexity (in that it is a means of drawing attention to the limitations of our models, and the implications that these limitations hold for our theories and practices), and the complexity of ethics (because, in opening the door to difference and otherness, and in disrupting the unity of our conceptual schemas, deconstruction highlights the fact that our models cannot fully capture or exhaust ethical concerns).

Deconstruction, as a mechanism for articulating ethical complexity, has an ambiguous status. Derrida (1988a) writes that it is impossible to conclusively define deconstruction as an analysis, a critique, a method, or even an act or operation (Derrida 1988a), since doing so, denies other understandings of the term, and thereby destroys the very complexity that deconstruction seeks to safeguard. In other words, in defining, we neutralise the force of a disruption or an interruption brought about by the very 'act' of deconstruction – hence, Derrida's (5) statement: 'What deconstruction is not? everything of course! What is deconstruction? nothing of course!'. Despite arguing that deconstruction represents a postmodern position, these definitional difficulties also problematise such a claim, because presupposed in such a claim is the notion that deconstruction can be identified with an existing body of knowledge or movement (albeit a highly fragmented one!).

Conceptualising deconstruction as a type of analytic strategy or method that can be unproblematically applied, as has often been the case in organisation studies (see Calás 1993; Boje 1998; Martin 1990), holds three additional dangers, as identified by the business ethicist and Derridean scholar, Campbell Jones (2004): firstly, there is the danger of reducing deconstruction to a mere method, thereby negating the epistemological, ontological, ethical, and political aspects of deconstruction; secondly, deconstructing can easily be construed as an activity conducted from outside the text ('a position of safe exteriority, if not objectivity' (41)); and, thirdly, deconstruction might be applied as a method by which to read a text, rather than a manner in which to renegotiate textual limits and relations (also see Jones 2003, 2007; Critchley 1999a). With these warnings in mind, a provisional description of deconstruction is nevertheless attempted in this chapter, followed by a discussion of the ethical implications that this view of deconstruction holds. As will be argued, a deconstructive ethics represents an ethically-complex position, and can therefore contribute to our understanding of both the ethics of complexity and the complexity of ethics.

Derrida's Central Concepts

Beyond Logocentricism

Hierarchy and Authority

Deconstruction works from the premise that in order for there to be meaning, reality must be interpreted and ordered (this insight is affirmed by the analysis on complexity, wherein it was stated that complex systems are structured systems that emerge over time due to dynamic self-organising processes). This ordering of reality into conceptual schemas creates hierarchies, where certain terms are necessarily privileged over others. The logos or privileged term operates at the expense of the marginal or secondary term, which is often employed to secure the status of the logos. Therefore, as Derrida (1981a: 41) explains in 'Positions', in any system of meaning 'we have not a peaceful coexistence of facing terms but a violent hierarchy. One of the terms dominates the other (axiologically, logically, etc.), occupies the commanding position.' The term which is in the commanding position – or in the authoritarian position – can only maintain its status by suppressing the secondary term, and by assimilating any differences into the logos.

It is precisely because of this assimilation of difference that Derrida takes exception with hierarchies and authority, as he explains in *Negotiations*:

> I have, it seems, a quasi-aesthetic aversion to authority and hierarchy... The aesthetic aversion has to do more with the fact that, most often, the most common forms of authority and hierarchy, of power and hegemony, have something in them which is vulgar, insufficiently refined, or insufficiently differentiated: thus my aversion to authority, in this case, is also an aversion to what is still too homogenous, insufficiently refined or differentiated, or else egalitarian (Derrida 2002a: 20).

However, Derrida is also quick to qualify that we cannot do without hierarchies and authority, as is clear from the following two citations: '... I do not think that there are nonhierarchical structures. I do not think they exist' (21); and '... I am not an enemy of hierarchy in general and of preference nor even of authority' (21). Hierarchies are the means by which we structure reality. In other words – and as argued in the previous chapter, hierarchies are a necessary condition for meaning; but – to reiterate – as soon as we have meaning it can (and should) be deconstructed, in order to guard against the naturalisation of these hierarchies. In destabilising hierarchies through deconstruction, we open up our systems of meaning in ways that allow for more differentiation and qualitative difference (21). For Robert Cooper (1989: 480), deconstruction therefore addresses 'the logics of *structure* and *process* and their interaction', by showing that 'our traditional ways of thinking are structure-biased and therefore incapable of revealing the nomadic and often paradoxical character of process.'

(Con)text

What makes the task of deconstruction difficult is that there is no neutral vantage point from which to deconstruct. To understand this statement it is necessary to explore in more detail the passage from *Of Grammatology* that was cited in the introduction. Derrida (1976: 158) states: 'There is no outside text' ['*Il n'y a pas de hors-texte*' (Derrida 1967: 227)] and 'There is nothing outside of the text' (Derrida 1976: 163) ['*Il n'y a rien hors du texte*' (Derrida 1967: 233)]. Critchley (1999a) notes the difference in formulations: the first citation claims that there is no 'outside-text' (which is an ontological point), whereas the second formulation claims that one cannot move outside textuality – there is nothing outside textuality (which is an epistemological point).

Despite the fact that we cannot escape textuality or the text, Derrida (1979: 81) argues in 'Living On: Border Lines' that 'no meaning can be determined out of context, but no context permits saturation'. This is due to the structure of meaning, which is informed by iterability, which Derrida (1988b: 7) describes in 'Signature Event Context' as 'the logic that ties repetition to alterity'. For meaning to be recognisable it needs to be repeatable (in other words, words must be understood across contexts). However, as soon as words enter a new context, their meanings change. Meaning is iterable in the sense that its communicability depends neither on the intentions of the author, nor on the context of its composition (7–9). Meaning disseminates, it moves out from the word resulting in a proliferation of meaning that cannot be controlled. In Woermann and Cilliers (2012: 452–453), the following example is provided to explain the iterable nature of meaning:

> A good illustration of this concerns how we understand the term 'freedom', used as the title of Jonathan Franzen's (2010) recent book, compared to its use in the title of Mandela's (1994) autobiography, 'A Long Walk to Freedom'. In both counts, the term 'freedom' is familiar, but in the former use it denotes a critical appraisal of contemporary American society; whereas in the latter use, the term is associated with the liberation struggle, and the story of Mandela's own imprisonment.

However, even supposing that we could fix meaning within a given context, we find that context is boundless: every context is open to further description, and meaning changes as the interpretation of the context changes. Again, Woermann and Cilliers (2012: 453) explain this point as follows:

> With regard to our example, one can argue that our understanding of 'freedom' in Mandela's autobiography is dependent on our own personal background. Not only does the understanding of the concept vary from one person to the next, but the very same book can never be reread in exactly the same way: what Mandela's autobiography, and the importance of freedom meant to me ten years ago, will differ from what it will mean if I were to reread the book today.

Any attempt to fix context through codification 'can always be grafted onto the context it sought to describe, yielding a new context which escapes the previous formulation' (Culler 1983: 123–124). Therefore, although we are embedded within the text (or, in complexity language, within a network of relations), the text itself is infinitely rich.

The structure and richness of textuality also undermines constructivist interpretations of Derrida work, which reduce all our experiences of the world to language. Rather, as David Schalkwyk (1997: 387) explains, both language and perception are regarded as 'species of the genus representation-in-general', or what Derrida sometimes calls the 'general text', but more often just 'the text'. Derrida therefore does not conflate language and the world, but 'insists on the imbrication of language and the world' (387), as he makes explicit in the 'Afterword', in stating that:

> The phrase which for some has become a sort of slogan, in general so badly understood, of deconstruction ("there is nothing outside the text: [*il n'y a pas de hors-texte*]"), means nothing else: there is nothing outside context. In this form, which says exactly the same thing, the formula would doubtless have been less shocking (1988c: 136).

In arguing that there is no Archimedean point that is 'out there' or independent of language, Derrida is also arguing against logocentrism, understood in this context as 'the determination of the being of the entity as presence' (Derrida 1976: 12). This is because logocentricism informs all 'names related to fundamentals, to principles, or to the center', all of which have 'always designated an invariable presence – *eidos, arche, telos, energia, ousia, aletheia*, transcendentality, consciousness, God, man, and so forth' (Derrida 1978: 279–280). Therefore, far from presenting an argument for constructivism (or the primacy of language), Derrida's philosophy works against logocentrism. A possible reason why critics continue to misunderstand Derrida's position on (con)text might be that he develops his most famous argument against logocentrism with reference to the opposition between speech and writing in language.

The Example of Speech and Writing

David Farrell Krell (1988: 8) explains that, traditionally, writing has always been characterised as materiality and exteriority, 'two explicitly excremental epithets'; whereas speech has been understood as the 'diaphanous, diaphonic ideality and interiority of the voice *qui s'entend parler* [who hears]'. The philosophical voice, who hears and understands itself, is a fully present voice, which represents the 'dream of totalising self-presence, perfectly fulfilled, utterly slaked desire' (9). Jonathan Culler (1983) argues that, due to this view, philosophical discourse has always defined itself against writing, and the threat posed by writing is that it would affect or infect the meaning it is supposed to represent. The ideal would be to contemplate thought directly; and, since this is impossible, we should strive for a language that is as transparent as possible and therefore as free of writing as possible. Culler further argues that in deconstructing the hierarchy that structures our understanding of language (as demonstrated further in the analysis), Derrida shows how logocentrism – or in this case, phonocentrism, as the privileging of voice – denotes a metaphysics of presence, which not only privileges speech before writing, but also:

> the positive before the negative, the pure before the impure, the simple before the complex, the essential before the accidental, the imitated before the imitation, etc. And this is not just one metaphysical gesture among others; it is the metaphysical exigency, that which has been the most constant, most profound, and most potent (Derrida 1988d: 93).

In summary, to say that there is nothing outside of (con)text is a way of asserting oneself against 'the Logos, the undeconstructable origin of the meaning of being, the rationality of thought, the absolute interiority of truth' (Lucy 2004: 71); or, of positioning philosophy inside the text. The question that now arises however is if there is nothing outside of con(text), how is one to open up texts or destabilise hierarchies? In other words, how does deconstruction take place?

On Deconstructing

The answer to the above question lies in the double movement of deconstruction. Deconstruction works on two fronts: on the one hand, to deconstruct means to concede to one's complicity in the systems of meaning which one seeks to challenge; and, on the other hand, it means to traverse beyond the system. These two movements of deconstruction do not follow chronologically, but take place *simultaneously*, and therefore require a double-thinking on the part of the deconstructionist.

The First and Second Movements of Deconstruction

The first movement of deconstruction demands that one engages with the dominant interpretation of a text or context. In order to successfully engage with a text, one must be competent at reading and writing so that 'the dominant interpretation of a text can be reconstructed as a necessary and indispensible layer or moment of reading' (Critchley 1999a: 24). If this were not possible, 'one could indeed say just anything at all' (Derrida 1976: 158), which is a strategy that Derrida explicitly renounces, in stating that '...I have never accepted saying, or being encouraged to say, just anything at all.'[1]

---

[1] This argument not only applies to the reading of certain texts (although Derrida was a great reader of texts, including the texts of Husserl, Hegel, Heidegger, Freud, Nietzsche and Saussure), but also to the reading of contexts. In this regard, it is useful to again recall Derrida's (1999: 67) views on close reading:

> I would assume that political, ethical and juridical responsibility requires a task of infinite close reading. I believe this to be the condition of political responsibility: politicians should read. Now to read does not mean to spend nights in the library; to read events, to analyze the situation, to criticize the media, to listen to the rhetoric of demagogues, that's close reading, and it is required more today than ever. So I would urge politicians and citizens to practice close reading in this new sense, and not simply to stay in the library.

It might seem strange that deconstruction – as a strategy employed against traditional structures of meaning – places such a significant emphasis on the dominant understanding of texts, but Derrida (1978) notes that all attempts at destructive discourses are trapped within a circle (and here he cites the Nietzschean critique of metaphysics; the Freudian critique of self-presence; and, the Heideggerean destruction of metaphysics, of onto-theology, and of the determination of Being as presence). This circle can be described as a performative contradiction, in which the destruction of the history of metaphysics can only be undertaken on the basis of this very history. In this regard, Derrida (280) writes:

> There is no sense in doing without the concepts of metaphysics in order to shake metaphysics. We have no language – no syntax and no lexicon – which is foreign to this history: we can pronounce not a single destructive proposition which has not already had to slip into the form, the logic, and the implicit postulations of precisely what it seeks to contest.[2]

However, despite these difficulties, a deconstructive reading 'cannot simply be that of commentary nor that of interpretation' (Derrida 1973a: 88). As stated before, to deconstruct means to destablise the dominant interpretation, whilst simultaneously engaging with this interpretation. The destabilisation of hierarchies should not be seen as a negative, or destructive enterprise. Rather, deconstruction (and destabilisation) is required for progress, for 'what remains to be thought beyond the constructivist or destructionist schema' (Derrida 1988c: 147).

The deconstructionist therefore takes the status quo seriously and works from within the existing system of meaning, in order to break it open to new interpretations. In the words of Derrida, deconstruction 'interven[es] in the field of oppositions it criticizes' (Derrida 1988b: 21). This is done so as to reveal the text's blind spot (or the repression and marginalisation of difference). This operation typically characterises the second movement of deconstruction (which, to reiterate, takes place alongside the first movement). In showing how that which is relegated to the margins of the dominant discourse or schema is needed to maintain the privileged status of the logos, the text starts to undo itself, thereby creating an opening for a new interpretation or way of being.

Revisiting the Example of Speech and Writing

In returning to the example of language, Derrida (1976) notes that if writing has always threatened the purity of speech, then the relationship between speech and writing is more complex than is portrayed in the traditional hierarchical schema, which gives precedence to speech over writing. In order to show how the hierarchical opposition between speech and writing can be reversed, Derrida begins his

[2] In this regard, it is useful to recall Alfred Whitehead's (1979: 39) famous claim that the 'safest general characterization of the European philosophical tradition is that it consists of a series of footnotes to Plato.'

deconstruction with reference to the Saussurian understanding of language as a system of differences (as introduced in Chap. 1). Recall that for Saussure (1960), there is no exact or literal meaning that is inherent to language – rather, all we have is a dynamic system of differences, where every event or every speech act is itself made possible by prior structures. Derrida appropriates the Saussurian insight that, instead of having a substantive understanding of words or concepts, our understanding of language is premised on how words and concepts differ from one another.

The notion of the sign, as a differential unit, is best illustrated in the written form, where, for example, we recognise the letter 'm' as distinct from the other letters in the alphabet, which allows us to form a relational understanding of the letter 'm'. In explaining the differential nature of speech, Saussure (1960: 119) writes: 'Since an identical state of affairs is observable in writing, another system of signs, we shall use writing to draw some comparisons that will clarify the whole issue.' Hence, as Culler (1983: 101) explains, we see that writing, which, for Saussure, should 'not be the object of linguistic enquiry, turns out to be the best illustration of the nature of linguistic units.' In this regard, Derrida (1976: 44) states:

> If 'writing' means inscription and especially the durable instituting of signs (and this is the only irreducible kernel of the concept of writing), then writing in general covers the entire domain of linguistic signs... The very idea of institution, hence of the arbitrariness of the sign, is unthinkable prior to or outside the horizon of writing.

In the second movement of deconstruction, the hierarchy of language is destabilised through reversing or overturning the binary opposition between speech and writing. This in turn gives us a new concept of writing, as a generalised writing (or arche-writing), which Culler (1983: 102, 101) argues is both 'the condition of speech and writing in the narrow sense', or which has as its subspecies 'a vocal writing and a graphic writing'. However, one should be careful not to immediately institute writing as the dominant term, as this will not result in a destabilisation of the hierarchy, but rather in the re-instutionalisation of another rigid structure. In this regard, Derrida warns that '[d]econstruction cannot be restricted or immediately pass to neutralization' (Derrida 1988b: 21). To avoid this, one should also be sensitive to the process of displacement or 'metaphorisation' (Cooper 1989: 483) of meaning that takes place in deconstruction.

Just as Derrida insists on the imbrication of language and the world; so too, the example of arche-writing points to the imbrication of speech and writing. In other words, the individual terms inhabit each other (Cooper 1989), or give way 'to a process where opposites merge in a constant *undecidable* exchange of attributes' (Norris 1987: 35). It is this process that guards against the naturalisation of hierarchies, since the undecidable exchange of attributes implies that meaning is constantly deferred. In order to better understand what is meant by the constant deferral of meaning, it is useful to turn to Derrida's discussion on the supplement, play, *différance*, and trace, which – as will be shown – are 'concepts' that are well illustrated by Derrida's discussion on Plato's pharmacy.

Beyond a Binary Logic

Supplementary Complications

As stated in the previous chapter, Derrida (1988c) reminds us in the 'Afterword' that the oppositional logic of hierarchies is necessary for us to generate concepts and meaning (we can have no non-hierarchical concepts). However, Derrida (117) also warns that 'the ideal purity of the distinctions proposed... is inaccessible' and that 'its practice would necessitate excluding certain essential traits of what it claims to explain or describe – and yet cannot integrate into the "general theory."' As such, 'all conceptual production appeals to idealization'. For this reason, deconstruction (as the destabilisation of hierarchies) is necessary to provide:

> a supplementary complication that calls for other concepts, for other thoughts beyond the concept and another form of "general theory," or rather another discourse, another "logic" that accounts for the impossibility of concluding such a "general theory" (117).

To understand what Derrida means by a supplementary complication, we turn to his reading of Rousseau in *Of Grammatology*. Derrida (1976: 163) writes that the word *supplément* is the 'blind spot' in Rousseau's texts, in that he employs the word without accounting for its logic. Culler (1983) explains that Rousseau uses the term supplement to describe the relationship between speech and writing. For him, writing is a technique which is foreign and exterior to speech, and which is therefore an add-on. However, writing can only function as a supplement to speech, if speech is not sufficient unto itself; or, in the words of Derrida (1976: 103), writing is a derivative of speech 'only on one condition: that the 'original,' 'natural' etc. language never existed, was never intact or untouched by writing, that it has itself always been a writing'.

The work of the supplement is to substitute for an absence or lack in the logos; but, as Derrida (1978: 289) notes in his discussion on supplementarity in the work of Levi-Strauss (1966), such an absence or lack is not something that we can recover through rigorous work, but is rather inherent to the nature of concepts: 'One cannot determine the center and exhaust totalization because the sign which replaces the center, which supplements it, taking the center's place in its absence – this sign occurs as a surplus, as a *supplement*' (Derrida 1978: 289). Absence therefore defines the very heart of the logos, which must be supplemented.

Play, *Différance*, and the Trace

The logic of supplementarity is always at play in our concepts. All systems of meaning can be conceptualised on a continuum between no centre (absolute free play) and a fixed centre (absolute structure). The concept of play (*jeu*) is employed by Derrida to denote the fact that no completeness or totalisation is possible. Derrida (1978) notes that, in the classical style, totalisation refers to a subject or finite richness that cannot be empirically mastered, simply because the subject matter lends itself to a conceptual richness that cannot be captured in a finite discourse. However, for Derrida, the impossibility of totalisation is not due to empirical limitations, but rather due to the nature of the subject or field:

> The field is in effect that of *play*, that is to say, a field of infinite substitutions only because it is finite, that is to say, because instead of being an inexhaustible field, as in the classical hypothesis, instead of being too large, there is something missing from it: a center which arrests and grounds the play of substitutions (289).

Derrida (290) proceeds in describing the function and nature of play: 'Play is the disruption of presence' and is 'always the play of absence and presence, but if it is to be thought radically, play must be conceived of before the alternative of presence and absence.' This description of play is remarkably similar to Derrida's (1981b: 27) description of *différance*, which he characterises as:

> a structure and a movement that cannot be conceived on the basis of the opposition presence/absence. *Différance* is the systematic play of differences, of traces of differences, of the spacing [*espacement*] by which elements relate to one another. This spacing is the production, simultaneously active and passive (the *a* of *différance* indicates this indecision as regards activity and passivity, that which cannot be governed and organized by that opposition), of intervals without which the "full" terms could not signify, could not function.

In French, the word for 'difference' is *différence*; however, the difference between *différence* (with an 'e') and Derrida's term *différance* (with an 'a') is inaudible. In other words, the identity of *différance* only exists in writing (Lucy 2004). *Différance* as 'the systematic play of differences' not only refers to the space between 'a' and 'e' (the spacing of difference), but also to the necessity of spacing, as the means by which elements are related to one another (spacing as difference). Niall Lucy (26) explains that whereas the 'spacing of difference' is a passive spacing, 'spacing as difference' constitutes an active movement in time. *Différance* is therefore both a spatial and a temporal concept, where the meaning of an element constantly differs (*diffère*) from the meaning of other elements, but where meaning and identity are also constantly deferred (*différer*). This means that identity is constituted by relational difference (Saussure's insight), but also that – because identity is constituted by difference – an element's "own" constitution as an autonomous or fully complete entity' (27) is always deferred. Derrida (1982: 13) writes that '[i]t is because of *différance* that the movement of signification is possible', which means that '*[d]ifférance* [like play] is neither a word nor a concept' (Derrida 1973b: 130) but rather the condition of possibility for conceptuality and words as such (Critchley 1999a).

Whereas binary and logocentric schemas emphasise the difference between opposing terms, deconstruction, supplementarity, play and *différance*, show how our signifying systems are constituted by a difference that both separates and joins. In so doing, attention is drawn to 'a rupture within metaphysics, a pattern of incongruities where the metaphysical rubs up against the non-metaphysical, that it is deconstruction's job to juxtapose as best as it can' (Reynolds 2005). Derrida marks this rupture by employing the term 'trace', which Spivak characterises as 'the mark of the absence of a presence, an always already absent present, of the lack at the origin that is the condition of thought and experience' (preface to *Of Grammatology*, Derrida 1976: xvii). The trace never appears as such, and has no meaning in itself, because it has 'no place, for effacement belongs to the very structure of the trace' (Derrida 1973b: 156). However, according to Jack Reynolds (2005), a deconstructive intervention can mime the logic of the trace in a text, and bring it to the fore. Meaning is dependent on the interaction of many traces, which also serve to prevent metaphysical closure and the naturalisation of hierarchies, which, as Derrida points out, is an important task, because 'one always inhabits, and all the more when one does not suspect it' (Derrida 1976, 24).

In layman's terms, the logic of play, *différance*, and the trace reveal how the meaning of our concepts are always-already contaminated by the meaning of other concepts, which serves to undermine the rigidity of our conceptual distinctions. In order to illustrate the nature of the trace, as well as the temporal or processual nature of *différance*, we turn to the example of the *pharmakon*.

Deconstructing Plato's Pharmacy

In the essay entitled 'Plato's pharmacy', Derrida (2004a) deconstructs several texts by Plato, including *Phaedrus*, in order to reveal the interconnections between *pharmakon* and *pharmakeus*, and the notable absence of the word *pharmakos*. The word *pharmakon* means both remedy/cure and poison. In *Phaedrus*, Plato (1997) uses the term to describe writing: writing is a supplement – the artificial add-on which both poisons and cures. Culler (1983) explains that in this text, Plato employs the term in a fashion that suggests that the meaning of the term (as either remedy or poison) is clear. However, what Derrida shows in 'Plato's pharmacy' is that the character of the *pharmakon* is undecidable and ambivalent. According to Derrida (2004a: 13), the *pharmakon* 'constitutes the medium in which opposites are opposed, the movement and the play that links them among themselves, reverses them or makes one side cross into the other'. In other words, one can never be entirely sure whether *pharmakon* refers to a poison or a cure, since traces of each meaning pervades the other. Cooper (1989: 489) elaborates on Derrida's understanding of the *pharmakon*, by writing that the two meanings of *pharmakon* 'actively *defer* each other, the deferred term being postponed for the present, waiting for an opportunity to flow back to the medium from which it was severed.' The *pharmakon* is therefore '(the production of) difference. It is the difference of difference' (Derrida 2004a: 130).

The logic of the *pharmakon* infects both speech and writing, by showing how the oppositional difference between these terms is unsustainable; or, by showing how the play of *différance* always precedes an oppositional difference. In this regard, Derrida (113) writes:

> Plato maintains *both* the exteriority of writing *and* its power of maleficent penetration, its ability to affect and infect what lies deepest inside. The *pharmakon* is that dangerous

> supplement that breaks into the very thing that would have liked to do without it yet lets itself *at once* be breached, roughed up, fulfilled, and replaced, completed by the very trace through which the present increases itself in the act of disappearing.

Culler (1983: 143) writes that 'this role of the *pharmakon* as a condition of difference is further confirmed by the link with *pharmakos*, "scapegoat".' Derrida (2004a) argues that, just as Socrates wished to exclude the poison of writing from the purity of speech, so too the exclusion of the scapegoat from the city is meant to represent the casting out of evil. In fact, the Greeks held a special purification and expiatory ceremony on the sixth day of the Thargelia (an Athenian festival in honour of Apollo and Artemis), in which two *pharmakoi* were chosen to die. However, Derrida makes the significant point that the evil of the *pharmakos* can only be defined from within the city walls, even though the aim is its 'exclusion out of the body (and out) of the city' (133). This leads him to conclude that:

> The ceremony of the *pharmakos* is, thus played out on the boundary line between the inside and the outside, which it has as its function ceaselessly to trace and retrace. *Intra muros/ extra muros*. The origin of difference and division, the *pharmakos* represents evil both introjected and projected (134).

Derrida notes that the word *pharmakos* is also a synonym for *pharmakeus*, which means 'wizard, magician, poisoner' (133). In Plato's dialogues, Socrates is often described as a *pharmakeus* – as a sorcerer – whose philosophy represents a *pharmakon*; a medicine or cure in the form of '[t]he eidos, truth, law, the episteme, dialectics [and] philosophy' (127). However, Socrates' *pharmakon* 'must be opposed to the *pharmakon* of the Sophists and to the bewitching fear of death' (127). As such, the Sophists are also portrayed as *pharmakeia*, but instead of healing, these witches prescribe poisonous sophistry. Therefore, as Derrida (127) notes, Plato pits '*pharmakeus* against *pharmakeus*, *pharmakon* against *pharmakon*.'

Ironically however, Socrates also becomes Athens' most famous scapegoat, and was eventually killed by ingesting poison. What is even more remarkable is that the date designated for the ceremony of the *pharmakos* (namely, the sixth day of the Thargelia) marks 'the day of the birth of him whose death – and not only because a *pharmakon* was its direct cause – resembles that of a *pharmakos* from the inside: Socrates' (135). Today, we can recall this incident due to its inscription in writing, which, to Socrates, represented 'a harmful drug [and] a philter of forgetfulness' (129) i.e. the most dangerous *pharmakon* of all!

The interplay of *pharmakon-pharmakos-pharmakeus* illustrates the play of meaning and of *différance*. In this regard, Derrida (128) notes that 'the essence of the *pharmakon* lies in the way in which, having no stable essence, no 'proper' characteristics, it is not, in any sense (metaphysical, physical, chemical, alchemical) of the word, a substance.' There is no sure way of keeping the poison and the cure, the magician and sorcerer, speech and writing, or logic and rhetoric apart. All these concepts bear the mark of the trace: their meanings are established by an endless chain of supplements, by the play of competing interpretations, and by relations and deferrals that are constantly at work.

Deconstruction Is Hymeneal

In summary, we can say that all meaning is characterised by the double movement, which like the hymen – i.e. that fold of tissue that covers the jewel box of virginity – simultaneously presents the membrane of meaning, and signifies a penetration of this membrane and meaning. The hymen does not belong to the inside or to the outside: it is the 'in-between'; it signifies the space of the partition, or the spatio-temporal interval between 'the presence of this' and 'the presence of that'. The hymen serves to de-ontlogise the centre, by ontologising the margin (between inside and outside), and is thus the movement of work which joins and separates, connects and divides (see Derrida 2004b; Culler 1983).

With regard to the first movement, Culler (1983) notes that deconstruction is hymeneal in the sense that, on the one hand, it marks the place of difference. As with a traditional reading which necessitates engagement with the dominant interpretation of texts, the hymen marks the acceptance of the distinctions between the surface features of a discourse and the underlying logic; between the empirical features of language and thought itself; and, between the inside of the system and the outside of the system.

These distinctions become prevalent in the second movement[3] of deconstruction, where attention is drawn to the text's blind spots, including the metaphors, apparently marginal features of the text, and the different meanings of the words in the text. This is done in order to reveal the inherent paradoxical logic of texts, and to set forth a reversal of the dominant interpretation. In this regard, Culler (146) writes: 'Derrida is not playing with words, he is betting with words, employing them strategically with an eye on larger stakes.' However, merely reversing the dominant interpretation is not enough: a displacement of meaning should also take place.

Deconstruction is hymeneal in that it destabilises concepts. By teasing out textual and linguistic configurations (as in 'Plato's pharmacy'), the deconstructionist puts into question the possibility of distinguishing with surety between the oppositional concepts and operations that govern texts. In this regard, it is useful to bear in mind that the hymen is etymologically related to sewing, weaving, or spinning. Meanings are intertwined: there is always a trace of alterity or difference that pervades our concepts. As such, the hymen is the space of difference itself, or difference within concepts – which also prevents concepts from closing in upon themselves, and thus save us from solipsism.

---

[3] Culler (1983: 146) refers to the first and second movement as the right-handed and left-handed operations of deconstruction. This classification works well, as – unlike the chronology implied in the terms 'first' and 'second' – the terms 'left' and 'right' draw attention to the simultaneous nature of the task at hand.

Complexity, Deconstruction, and Ethics

The Ethics of Deconstruction

Ethical Testimony

In recent years, there has been much discussion regarding what appears to be an 'ethical turn' in Derrida's later works. Jones (2003) argues that whereas some see the ethical turn as constituting a break with the themes that dominate his earlier works, others maintain that the more explicit focus on ethics 'is nothing but a clarification or extension of themes ever present in deconstruction' (224). During the interview, 'Hospitality, justice and responsibility: a dialogue with Jacques Derrida', Derrida (1999) is asked a question regarding the continuity in his thinking, as the questioner is unconvinced that there is indeed something like a '*Kehre* or Heideggerean turning' in his thought. In response to this question, Derrida (81) states 'I am grateful that you don't want to cut me in two; I do wish to be cut, but in more than two places!' Derrida wrote on many themes and published extensively during his lifetime, and, as such, one should be weary of identifying an explicit turn in his work. Despite his sheer output, one can provide a stronger reason against a turn, namely that Derrida has been writing on ethics (even if it is not the only thing he has been writing on), since coining the term deconstruction.

During the above-mentioned interview, Derrida is asked whether there is 'a logic of ethical testimony at work in deconstruction?' To this he answers:

> Yes, it is absolutely central to it. Testimony, which implies faith or promise, governs the entire social space. I would say that theoretical knowledge is circumscribed within this testimonial space. *It is only by reference to the possibility of testimony that deconstruction can begin to ask questions concerning knowledge and meaning* (82; my italics).

Ethical testimony is the impetus for deconstruction. Without the promise of accounting for that which is excluded from our systems of meaning, or without deconstructing in the name of ethics or the Other, deconstruction would be an endless and pointless exercise. François Raffoul (2008: 272) cites a *Humanité* interview, in which Derrida (2004c; trans. F. Raffoul) states that a slogan of deconstruction is 'being open to what comes, to the to-come, to the other', and included in this openness is the question of ethics itself, since the very possibility of the ethicity of ethics is problematised in the *aporetic* thinking that characterises deconstruction. In the *Humanité* interview cited above, Derrida (2004c; trans. F. Raffoul) further states that in his work 'ethical questions have always been present, but if by ethics one understands a system of rules, of moral norms, then no, I do not propose an ethics'. The logic of ethical testimony at work in Derrida's writing should be understood in terms of 'a genuine philosophical questioning concerning the meaning of the ethical' (Raffoul 2008: 271), which Raffoul reiterates is necessarily informed by an *aporetic* thinking, an ethics of alterity, and of the welcoming of the other.

Ethics as the Unspeakable Limit of the World

In a very real sense, this ethics of alterity is impossible, because it denotes an attempt to think the outside of a system. However, a central premise of this study is that ethics, understood as corresponding with some transcendental realm, remains inaccessible to us, since that which lies outside of the system cannot be known in its alterity inside the system. William Rasch (2000: 80) writes that, in operationally-closed systems (i.e. systems that cannot access the environment except through doubling the distinction between the system and the environment inside the system itself), the outside serves 'as the absolute condition for the existence of the inside, but it remains supremely unknowable. It is the silence that delimits the world.' Given this understanding, ethics 'serves as the unspeakable limit or condition of the world' (77); or, in Ludwig Wittgenstein's (2001: Tractatus 6.421) famous words: '*Es ist klar, dass sich die Ethik nicht aussprechen lasst*' (It is clear that ethics cannot be articulated).

For Luhmann (2000), this means that all knowledge (including ethical knowledge) is the outcome of reflexive processes. Ethics, on his interpretation, cannot be employed to generate normative judgements; rather, it can only function as a historical description of the pragmatically-determined standards that inform our practices and language games. A problem that arises when we view complex systems as operationally-closed is that our language games and practices can bear solipsistic implications. In other words, if the ethical task is viewed only in terms of producing thicker descriptions of our practices and language games, we risk developing an incommensurable and relativist view of ethics, and moreover risk developing constructivist accounts of reality. However, Derrida (2002b: 367) clearly states that his ethics of alterity, which accounts for 'differences of every order, beginning with the differences of contexts', does not constitute an empiricist or relativist stance. Rather, Derrida's deconstructive enterprise constitutes an attempt to address the methodological complexity of thinking together a system and its environment; and, in so doing, he safeguards the possibility of an ethics that cannot be appropriated, whilst simultaneously avoiding the reinstatement of metaphysical ideals.

Derrida's Limit Concepts

Derrida's attempt at thinking transcendental ethics within the world is made possible by his quasi-transcendental or limit concepts, of which there are numerous, but which, perhaps most famously, include the idea of justice, the gift, and hospitality. Gunther Teubner (2001: 42; my italics) brilliantly captures the methodological complexity involved in thinking through these limit concepts, with specific reference to justice and the gift:

> This is Derrida's central thesis: justice as transcendence in an irreconcilable contrast to the immanence of positive law which, however, is haunting law constantly. And it is here that the parallels to the relationship of gift to circulation of the economy become visible. The gift is not only... an ethical or political counter-principle to the cold economic logic of capitalism. The gift transcends every social relation and provokes it... as an unconditional demand for the Other (Derrida 1992: Ch. 2). *The gift is not pure transcendence without any*

> *connection to the circulation to the economy, but in a self-contradictory relation of unrevisable separation and permanent provocation.* Thus Derrida's repeated calls for political engagement, which are provoked by the unsatiated demands of a transcendent justice and gift (Derrida 1990: 933).

Although justice can never be encapsulated in law, and although the gift can never be given within the framework of the economy, these limit concepts, nonetheless, shape and drive the ethics of our institutions. Teubner (2001: 43) argues, with reference to the practice of law, that it makes an important difference whether or not 'one is exposed to a 'deconstructive justice', to extreme demands of justice that can never be realized… to a sense of fundamental failure of law, even a tragic experience that whatever you decide in law will end in injustice and guilt.' In other words, we must judge; however, we cannot pass judgement without being acutely aware of the fact that we have failed justice.

Thinking through limit concepts therefore demands that we simultaneously account for a system and its environment, or for the known and the unknown. This demands a measure of complex thinking, because, to recall the words of Morin (2008: 11), within open systems '[t]he environment is at the same time intimate and foreign: it is part of the system while remaining exterior to it'. In other words, the system and its environment are *aporetically* related to each other. Understanding and engaging with this relation, in part, defines the challenge of deconstruction: on the one hand, we cannot completely renounce our tools and systems of meaning; but, on the other hand, it is exactly these tools and systems (including language itself) that should be challenged, if we hope to act justly.

In this regard, Derrida (2002b: 364) writes that '[t]he deconstruction of logocentricism, of linguisticism, of economism… etc., as well as the affirmation of the impossible are always put forward in the name of the real, of the irreducible reality of the real'. Derrida (264) warns that it is important not to think about the real as an 'attribute of the objective, present, perceptible or intelligible thing (res)'. Rather, for him, the real should be understood as a promise, or 'as the coming or event of the other, where the other resists all reappropriation, be it ana-onto-phenomenological appropriation' (367).

The nature of this argument becomes clearer if we consider the following remarks on forgiveness, which – like justice, hospitality and gift-giving – has the same structure as ethical decision-making. Derrida (2002b: 351) writes that true forgiveness cannot stem from duty: 'One forgives, if one forgives beyond any categorical imperative, beyond debt and obligation.' In other words, forgiveness must be unconditional. It must disavow the tradition of repentance, economic exchange, and identification, even though we can only think about forgiveness within this tradition.

The Impossible Possibility of Ethics

Deconstruction, as the means by which to articulate the ethical interruption of ontological closure (as expressed in limited concepts such as unconditional hospitality, gift-giving, and justice), always operates in service of the singularity of the event, which both calls for an 'appropriative reception' (Raffoul 2008: 275), but in which the appropriation necessarily fails. This is because after the decision is taken and the action assumed, the ethical moment is transformed into something that can be defined, justified, and criticised. In other words, the promise is ruined because the logic of the event is destroyed, assimilated into a generalised hierarchy of meaning that transforms ethics and justice into codes and law, and that is therefore once again open to deconstruction. In this regard, Derrida (2003: 90) writes that 'there is no event worthy of its name except insofar as this appropriation falters [*échoue*, literally: fails] at some border or frontier'.

Derrida's entire body of work therefore serves as a constant reminder of the necessity of remaining open to an alterity that cannot be assimilated, but that nevertheless demands our consideration. Derrida (2004c; trans. F. Raffoul) urges us to think on the impossible, and notes that '[t]o do the impossible cannot be an ethics and yet it is the condition of ethics. I try to think the possibility of the impossible.' For Derrida, 'possibility' and 'impossibility' are not modal opposites; rather, the impossibility of ethics, serves as the condition for its very possibility. This point is easier to comprehend in the language of complexity: deconstruction is a means of trying to think through the complexity that precedes the reductive implications introduced by modelling. This is, of course, impossible since we cannot understand phenomena in their full complexity, but even though this complexity escapes our understanding, acknowledging it already implies an engagement with the demands imposed on us by this complexity. Returning to Derrida's understanding of the possible and the impossible, Raffoul (2008: 273) writes:

> The impossible would no longer be the opposite of the possible but, on the contrary, would be what "haunts the possible," [Derrida 2001: 98; trans. F. Raffoul] what truly "enables" or possibilizes the possible. The impossible, Derrida would claim, is possible, not in the sense that it would become possible, but in a more radical sense in which the impossible, as impossible, is possible.

Therefore, on the one hand, it is impossible to understand ethics, which presents itself as an experience of the impossible or the limits of the world (which is Wittgenstein's point); but, on the other hand, this impossibility is also what serves as the very condition for the possibility of ethics (Raffoul 2008), which finds its expression in the moment of the event. The *aporia* of ethics (i.e. the possibility of its impossibility) requires endurance in thinking, and we should resist the desire to halt at this *aporia* or to overcome it (Derrida 1993).

Implications

Deconstruction, as a project aimed at safeguarding difference, constitutes an engagement with the ethics of complexity. More specifically, Derrida's philosophy presents a way of pulling the insights gleaned from an account of critical complexity into the human domain, and thereby provides us with a non-trivial and productive reading of the ethics of complexity. This is because deconstruction offers a radical challenge to the models or conceptual schema that we employ to make sense of our complex realities, including our ethical models and the models that we draw of ourselves.

Ethical Models: The State of 'Applied' Ethics

With specific reference to the current state of applied ethics, Raffoul (2008: 271) writes that:

> It is indeed a remarkable and odd anomaly of the contemporary philosophical field that the professional philosophers of ethics, the so-called 'ethicists,' do not in general raise the question of the ethicality of ethics, too busy that they are to "apply" it, as they say. In such a context, where the current and growing development of "applied ethics" in the curriculum is paradoxically accompanied by a peculiar blindness regarding the nature of ethics... and the meaning of the ethical, it would be crucial to raise anew the question or questions on the meaning of ethics.

Deconstruction (along with Derrida's emphasis on the supplementary complications, play, *différance*, and traces that pervade our conceptual schemas) is an important means of raising anew the questions on the meaning of ethics, which, as Derrida (1995a: 16) writes, 'still remains open, suspended, undecided, questionable even beyond the question'.

Again, with reference to applied ethics, we can say that this task is indispensible, given the fact that our models hold important ethical-political implications that affect the lives of people – either positively, if one's interests are accounted for in the espoused ideology; or negatively, if one's interests fall outside the domain of the mainstream ideology. Today, the dominant ideology in the global context is mostly still connected to the ideals of the European Enlightenment tradition, and all the associated predicates related to these ideals, including the history of metaphysics, the history of Being, and the history of the West (Derrida 1976). In a practical context, we see how these ideals shape policy issues, and serve to force other ways of being into the Western mould, which masquerades as the global standard. It is these types of concerns that are highlighted in the practice of deconstruction, and it is through raising these concerns that deconstruction keeps the question of ethics alive, open, and urgent.

Models of the Subject: The Differentially-Defined Subject

Deconstruction does not only shed light on the ethics of complexity, but also contributes to our understanding of the differential nature of agency and identity. In the previous chapter, the point was made that we co-constitute one another and our practices, which is an activity defined as an emergent process of becoming that takes place within a spatial network of relations. Derrida (1995b) radicalises this account of identity formation, in arguing that our identities are marked by traces and by *différance*, and therefore our understanding of ourselves as autonomous or fully constituted entities is always deferred. In other words, we are marked from the inside by irreducibility. This also means that there is no singular oppositional limit between myself and the other, or between the human and the infrahuman.

This insight holds important consequences for, what Derrida (285) terms, the 'ethics and politics of living' as is evidenced by the following argument and examples:

> There is no need to emphasize that this question of the subject and of the living "who" is at the heart of the most pressing concerns of modern societies, whether they are deciding birth or death, including what is presupposed in the treatment of sperm or the ovum, surrogate mothers, genetic engineering, so called bioethics or biopolitics (what should be the role of the State in determining or protecting a living subject?), the accredited criteriology for determining, indeed for "euthanastically" provoking death... organ removal and transplant... (283).

Derrida raises these examples, in order to show that '[w]e know less than ever where to cut – either at birth or death. And this also means that we never know, and never have known, how to *cut up* a subject. Today less than ever' (285). That we are less certain of how to deal with the question of the subject today is not only due to the fact that our identities are fluid, but is also an outcome of advances in all the sciences. This is because as they develop they create political problems, which, according to Morin (2008), means that politics has become very complex, since it is now concerned with all dimensions of humanity. However, Morin argues that the response to this complexity is dominated by economic and technical thinking, rather than forms of thinking capable of understanding politics in its multi-dimensionality.

In order to draw attention to the ethical implications that arise from the inextricable ways in which we are tied together, Derrida (1995b) introduces the metaphor of eating, which represents an attempt to 'determine the best, most respectful, most grateful, and also most giving way of relating to the other and of relating the other to the self' (281–282). The metaphor of eating highlights the manner in which we asymmetrically appropriate one another, both in terms of physical nourishment (eating, suckling) and in terms of our being-together-with-the-other in the world (in the sense of seeing, hearing, feeling, tasting, touching, and talking).

For Derrida (475, 482), our relations with one another should be governed by the ethical mandate: '*Il faut bien manger*', which should be understood both as 'one must eat well' (in the sense of 'learning and giving to eat, learning-to-give-the-other-to-eat'), as well as 'everybody has to eat'. And, the adverb 'bien' should be nominalised into 'Le Bien', to imply 'the eating of the Good.' Furthermore, because one never eats entirely on one's own, this ethical mandate is a rule offering infinite hospitality.

In drawing attention to the many ways in which we relate with one another, Derrida's ethical mandate shifts our understanding of responsibility as a causal process (in which we are accountable for our actions) to an understanding of responsibility that is grounded in a large explicit and tacit network of relations (in which we are accountable towards one another).

Deconstruction: A Negative Ethics vs. a Critical Ethics

Derrida's circumscription of deconstruction and of ethics as an experience of the impossible has sparked much controversy. Indeed, one need only take note of the critical obituaries written in the popular press after his death (see Kandell 2004; Mendez-Opale 2004), or the vehement protests voiced by a number of analytic philosophers at Cambridge, who tried to stop the university from awarding Derrida an honourary degree (see Smith 2005), in order to get a feel for the type of reactions that Derrida's deconstructive philosophy elicits. The following example taken from an article by Stephens (1994), serves to illustrate the malicious nature of these attacks. In this article, Roger Kimball (2008), a conservative critic and author of *Tenured Radicals*, is quoted as saying: 'Derrida's influence has been disastrous... He has helped foster a sort of anaemic nihilism, which has given imprimaturs to squads of imitators who no longer feel that what they are engaged in is a search for truth, who would find that notion risible.' Whilst some of his ardent followers have missed the point of deconstruction, it should be clear at this juncture that it is unfair to equate Derrida's project with nihilism.

All too often, deconstruction is construed as a negative ethics, intent on destroying rather than building-up. Such criticism confuses a critical ethics with a negative ethics, and is mostly propagated by those who wish to perpetuate the dream of a categorically-binding ethics. Derrida certainly aims to challenge conventional understandings of ethics, and in this sense he is critical of the Western tradition of philosophical thought. However, his criticisms are not intended to steer us towards the abyss of nothingness, but towards assuming a deeper responsibility for our decisions and actions, which is attained through grappling with the complexities of our lived experiences. To elaborate in the words of Derrida (1978: 292): a deconstructive ethics necessitates that we turn away from the 'dreams of deciphering a truth or an origin which escapes play and the order of the sign, and which lives the necessity of interpretation as an exile.' We need to turn away from moral 'recipes' that claim to lead us to the heart of ethics, and instead examine the margins, in an attempt to account for that which is excluded from these moral recipes. For this reason, Derrida writes that deconstruction – like negotiation – is also the very place of threat: 'one must [*il faut*] with vigilance venture as far as possible into what appears threatening and at the same time maintain a minimum of security – and also an internal security not to be carried away by this threat' (Derrida 2002a: 16–17). This threat is constituted by the fact that, in acting justly, we must renounce our systems of meaning, including consciousness, presence, and even language, even though we cannot do without these concepts. The challenge is to face the uncertainty, and to accept the responsibility that we bear for a future that is inherently characterised by risk.

The reason why the future is risky is because the product of our deliberations and actions can never be determined in advance, even though we are responsible for the consequences of our decisions. This is a daunting prospect, so daunting, in fact, that Derrida (1978: 293) argues that we tend to avert our eyes 'when faced by the as yet unnameable which is proclaiming itself and which can do so, as is necessary whenever a birth is in the offering, only under the species of the nonspecies, in the formless, mute, infant, and terrifying form of monstrosity.' In other words, despite our best intentions, our decisions and actions might turn out to be a mistake and we may open the door to a monster, but we can never know this in advance, and we need to risk it. The reason for this is that it is the only way in which to remain open to the Other and to the future, and to avoid 'end[ing] up in the well-managed dystopia of the brave new world' (Cilliers 2005: 264), in which nothing is ever risked and nothing is ever gained.

For some, however, the deconstructionist perspective presents a threat that is too large, and a promise that is too elusive, and it is this interpretation of Derrida's project that has often given rise to criticism. In this regard, Derrida (1988c: 140) states: 'I have come to understand that, sometimes, certain bitter and compulsive enemies of deconstruction stand in a more vital relationship, even if not theorized, to what is in effect at stake in it than do certain avowed "deconstructionists".' It is, ironically, precisely this recognition of what is at stake that may lead to the perpetuation of the status quo, because, as with complexity, taking these deconstructive insights seriously, poses a challenge to the whole system of thought that defines our views of our practices, politics, and ethics.

Deconstruction and Business Ethics

Despite the controversy unleashed by Derrida's insights, various authors have, since the early 1980s, sought to incorporate Derrida's ideas in the field of organisation studies (see Jones 2004: 34–35). A perfunctory glance at the organisational literature seems to suggest that, although popular in the 1980s and 1990s, Derrida's reception in the field has been short-lived. This is because it is generally believed that we have moved past postmodernism (and therefore past Derrida). In comparison to organisation studies, Derrida has not enjoyed much of a reception in business ethics. However, contrary to developments in organisation studies, we see that the incorporation of Derridean philosophy in business ethics has been more recent (see Jones 2003, 2004, 2007; Jones et al. 2005; MacKenzie 2000; Willmott 1998). Indeed, a special issue of *Business Ethics: a European Review* (vol. 19; issue 3) on Derrida, business, and ethics appeared in July 2010, following a conference held in 2008 on 'Derrida and Business Ethics'.

However, as in organisation studies (and philosophy in general), Derrida's reception in business ethics has – at best – been mixed. Given the above analysis of a deconstructive ethics, it is not surprising that many business ethicists (who concern themselves with legitimising the ethics of business, with providing ethical criteria against which organisations can measure their ethical success, or with engaging in business ethics management practices) would react with hostility to an ethics which speaks of deconstruction, supplement, play, trace, *différance*, and the (im)possibility of ethics! This hostility may, in part, stem from the fact that, in espousing codes, rules, norms, and procedures, these ethicists wish to offer comfort in the name of business ethics, by offering tools to 'solve' moral dilemmas. Against this, a deconstructive ethics has as its goal the task of calling into question these very tools that serve to '*produce* a reassuring sense of comfort' (Jones 2003: 238).

Yet, there is a second reason why many business ethicists are weary of bringing Derrida to business ethics, namely: the problem of application. Indeed, as Jones argues, merely the idea of applying Derrida already seems foreign to his thought, and the act of assimilating his ideas into a moral code or formula would destroy the very responsibility that such ideas seek to promote. In other words, deconstruction's anti-foundationalism proves a problem for application. Jones argues that as a result of this – and in an attempt to avoid the pitfall of moralising Derrida's ideas – business ethicists have often tended to speak in vague generalities about both postmodernism and a 'Derridean ethics', rather than pay attention to the specific arguments contained in Derrida's works. These vague generalities have limited the appeal and usefulness of incorporating Derrida's insights into business ethics. Richard De George (2008) explicitly conveys his skepticism regarding the purported value that Derridean insights offer for business ethics in arguing that:

> The onus is on Jones and other followers of Derrida to show how, by using 'the categories made available in the writings of Jacques Derrida' (Jones 2007), those in CSR and business ethics can do, and do more effectively, what they want to do and what they cannot do without these categories.

In Part II of this study, the theoretical basis developed in Part I is employed, in order to address DeGeorge's challenge. The application of this normative position to business ethics is undertaken, in order to show how a reading of a deconstructive ethics as a complex ethics can indeed lead to different – and possibly more useful – conceptualisations of prominent business ethics themes. However, before turning to the application, we conclude the first part of the study with an overview of the type of operations (or 'virtues') that support the theoretical position developed here.

References

Boje, D. 1998. Nike, Greek goddess of victory or cruelty? Women's stories of Asian factory life. *Journal of Organizational Change Management* 11(6): 461–480.

Calás, M.B. 1993. Deconstructing charismatic leadership: re-reading weber from the darker side. *Leadership Quarterly* 4: 305–328.

Cilliers, P. 2005. Complexity, deconstruction and relativism. *Theory, Culture & Society* 22(5): 255–267.

Cooper, R. 1989. Modernism, postmodernism and organizational analysis 3: The contribution of Jacques Derrida. *Organization Studies* 10(4): 479–502.

Critchley, S. 1999. *The ethics of deconstruction: Derrida and Levinas*, 2nd ed. Edinburgh: Edinburgh University Press.

Culler, J. 1983. *On deconstruction: Theory and criticism after structuralism*. London: Routledge.

De George, R.T. 2008. An American perspective on corporate social responsibility and the tenuous relevance of Jacques Derrida. *Business Ethics: A European Review* 17(1): 74–86.

Derrida, J. 1973a. Speech and phenomena: Introduction to the problem of signs in Husserl's phenomenology. In *Speech and phenomena and other essays on Husserl's theory of signs*, trans. D.B. Allison, 1–106. Evanston: North Western University Press.
Derrida, J. 1973b. Difference. In *Speech and phenomena and other essays on Husserl's theory of signs*, trans. D.B. Allison, 129–160. Evanston: North Western University Press.
Derrida, J. 1976. *Of grammatology*, trans. G. Spivak. Baltimore: Johns Hopkins University Press./ Derrida, J. 1967. *De la grammatologie*. Paris: Minuit.
Derrida, J. 1978. Structure, sign, and play in the discourse of the human sciences. In *Writing and difference*, trans. A. Bass, 278–294. Chicago: University of Chicago Press.
Derrida, J. 1979. Living on borderlines, trans. J. Hulbert. In *Deconstruction and criticism*, ed. H.I. Bloom, 75–175. New York: Continuum.
Derrida, J. 1981a. Positions. In *Positions*, trans. A. Bass, 37–96. Chicago: University of Chicago Press.
Derrida, J. 1981b. Semiology and grammatology. In *Positions*, trans. A. Bass, 15–36. Chicago: University of Chicago Press.
Derrida, J. 1982. Différance. In *Margins of philosophy*, trans. A. Bass, 1–28. Chicago: University of Chicago Press.
Derrida, J. 1988a. Letter to a Japanese friend. In *Derrida and différance*, ed. D. Wood and R. Bernasconi, 1–5. Evanston: Northwestern University Press.
Derrida, J. 1988b. Signature event context. In *Limited Inc*, ed. G. Graff and trans. S. Weber, 1–24. Evanston: Northern Western University Press.
Derrida, J. 1988c. Afterword. In *Limited Inc*, ed. G. Graff and trans. S. Weber, 111–160. Evanston: Northern Western University Press.
Derrida, J. 1988d. Limited Inc a b c.... In *Limited Inc*, ed. G. Graff and trans. S. Weber, 29–110. Evanston: Northern Western University Press.
Derrida, J. 1990. Force of law: The mystical foundation of authority. *Cardozo Law Review* 11: 919–1046.
Derrida, J. 1992. 'Counterfeit money' II: Gift & countergift, excuse & forgiveness. In *Given time: 1. Counterfeit money*, trans. P. Kamuf, 108–173. Chicago: University of Chicago Press.
Derrida, J. 1993. *Aporias*. Stanford: Stanford University Press.
Derrida, J. 1995a. Passions: An oblique suffering. In *On the name*, ed. T. Dutoit, 3–34. Stanford: Stanford University Press.
Derrida, J. 1995b. "Eating well," or the calculation of the subject. In *Points... Interviews, 1974–1994*, ed. E. Weber and trans. P. Kamuf, 255–287. Stanford: Stanford University Press.
Derrida, J. 1999. Hospitality, justice and responsibility: A dialogue with Jacques Derrida. In *Questioning ethics: Contemporary debates in philosophy*, ed. R. Kearney and M. Dooley, 65–83. London/New York: Routledge.
Derrida, J. 2001. *Dire l'événement, est-ce possible?* (with Gad Soussana and Alexis Nouss). Paris: L'Harmattan.
Derrida, J. 2002a. Negotiations. In *Negotiations: Interventions and interviews, 1971–2001*, ed. and trans. E. Rottenberg, 11–40. Stanford: Stanford University Press.
Derrida, J. 2002b. As if it were possible, "within such limits".... In *Negotiations: Interventions and interviews, 1971–2001*, ed. and trans. E. Rottenberg, 343–370. Stanford: Stanford University Press.
Derrida, J. 2003. *Philosophy in a time of terror*. Chicago: University of Chicago Press.
Derrida, J. 2004a. Plato's pharmacy. In *Dissemination*, ed. and trans. B. Johnson, 67–186. London/New York: Continuum.
Derrida, J. 2004b. The double session. In *Dissemination*, ed. and trans. B. Johnson, 187–316. London/New York: Continuum.
Derrida, J. 2004c. 'Jacques Derrida, penseur de l'évènement' [Jacques Derrida, thinker of the event], interview by Jérôme-Alexandre Nielsberg, *l'Humanité*, 24 Jan 2004. Available online at: http://www.jacquesderrida.com.ar/frances/evenement.htm.
Franzen, J. 2010. *Freedom*. Toronto: HarperCollins Canada.

Jones, C. 2003. As if business ethics were possible, 'within such limits'.... *Organization* 10(2): 223–248.
Jones, C. 2004. Jacques Derrida. In *Organization theory and postmodern thought*, ed. S. Linstead, 34–63. London: Sage.
Jones, C. 2007. Friedman with Derrida. *Business and Society Review* 112(4): 511–532.
Jones, C., M. Parker, and R. ten Bos. 2005. *For business ethics*. Oxon: Routledge.
Kandell, J. 2004. Jacques Derrida, abstruse theorist, dies at 74. *The New York Times*, 10 Oct 2004. Available online at: http://www.nytimes.com/2004/10/10/obituaries/10derrida.html?_r=1
Kimbal, R. 2008. *Tenured radicals: How politics has corrupted our higher education*, 3rd ed. Chicago: Ivan R. Dee.
Krell, D.F. 1988. Engorged philosophy: A note on Freud, Derrida and différance. In *Derrida and différance*, ed. D. Wood and R. Bernasconi, 7–11. Evanston: Northwestern University Press.
Levi-Strauss, C. 1966. *The savage mind*, trans. J. Weightman and D. Weightman. Chicago: University of Chicago Press.
Lucy, N. 2004. *A Derrida dictionary*. Malden/Oxford/Victoria: Blackwell.
Luhmann, N. 2000. Why does society describe itself as postmodern? In *Observing complexity: Systems theory and postmodernity*, ed. W. Rasch and C. Wolfe, 35–49. Minneapolis/London: University of Minnesota Press.
MacKenzie, R.N. 2000. Language, self, and business ethics. *Journals of Markets & Morality* 3(1): 22–42.
Mandela, N. 1994. *Long walk to freedom: The autobiography of Nelson Mandela*. London: Little, Brown.
Martin, J. 1990. Deconstructing organizational taboos: The suppression of gender conflict in organizations. *Organization Science* 1(4): 339–359.
Mendez-Opale, A.G. 2004. Jacques Derrida, French intellectual. *The Economist*, 21 Oct 2004. Available online at: http://www.economist.com/printedition/displayStory.cfm?Story_ID=3308320.
Morin, E. 2008. *On complexity*, trans. S.M. Kelly. Cresskill: Hampton Press.
Norris, C. 1987. *Derrida*. London: Fontana Press.
Plato. 1997. Phaedrus, trans A. Nehamas and P. Woodruff. In *Plato: Complete works*, ed. J.M. Cooper and D.S. Hutchinson, 506–556. Indianapolis: Hachett Publishing Company.
Raffoul, F. 2008. Derrida and the ethics of the impossible. *Research in Phenomenology* 38: 270–290.
Rasch, W. 2000. Immanent systems, transcendental temptations, and the limits of ethics. In *Observing complexity: Systems theory and postmodernity*, ed. W. Rasch and C. Wolfe, 73–97. Minneapolis/London: University of Minnesota Press.
Reynolds, J. 2005. Jacques Derrida. In *Internet encyclopedia of philosophy (IEP): A peer-reviewed academic resource*. Available online at: http://www.ieputm.edu/derrida/#SH3d. Last updated on 31 July 2005.
Saussure, F. 1960. *Course in general linguistics*, ed. C. Bally and A. Sechehaye, trans. W. Baskin. London: Peter Owen.
Schalkwyk, D. 1997. What does Derrida mean by 'the text'? *Language Sciences* 19(4): 381–390.
Smith, J.A. 2005. *Jacques Derrida: live theory*. London/New York: Continuum.
Stephens, M. 1994. Jacques Derrida. *New York Times Magazine*, 24 Jan 1994. Available online at: http://www.nyu.edu/classes/stephens/Jacques%20Derrida%20-%20NYT%20-%20page.htm.
Teubner, G. 2001. Economics of gift – positivity of justice: The mutual paranoia of Jacques Derrida and Niklas Luhmann. *Theory, Culture & Society* 18(1): 29–47.
Whitehead, A.N. 1979. *Process and reality*. New York: The Free Press.
Willmott, H. 1998. Towards a new ethic? The contributions of poststructuralism and posthumanism. In *Ethics and organizations*, ed. M. Parker, 76–121. London: Sage.
Wittgenstein, L. 2001. *Tractatus logico-philosophicus*, trans. D. F. Pears and B. McGuinness. London/New York: Routledge.
Woermann, M. and Cilliers, P. 2012. The ethics of complexity and the complexity of ethics. *South African Journal of Philosophy*, 31(2): 447–463.

Chapter 4
'Virtues' for a Complex World

Abstract In this chapter, which serves as the conclusion to the first part of the study, three operations are introduced, which help us to practically engage with the demands of the deconstructive and provisional account of ethics developed in this study, and which also serve to draw attention to 'the logic of complexity' that contaminates all conceptual schemas. These three operations, which have been termed 'virtues' for a complex world, are: transgressivity (which prevents us from simply reinforcing that which is current), irony (which allows us to recognise and engage with the limitations of a binary logic), and moral imagination (which allows us to successfully engage in critical meaning-producing processes that takes place within specific contexts defined by power and politics). The arguments put forward in this chapter are illustrated at the hand of examples from the organisational context, and from the management sciences and business ethics literature.

Introduction

Both the interpretations given in this study to complexity thinking and to Derrida's deconstructive philosophy underscore a critical position which, to recall, highlights the point that there is no position of safe exteriority from which we can practice a rationality or ethics premised on objectivity and universality. This, in turn, reinforces the point that all normative claims are provisional claims, and should be subjected to criticism, revision, and transformation (as implied both by the logic of the provisional imperative and deconstruction). Practically, it is very difficult to commit to a deconstructive or recursive view of ethics, since acknowledging that we are always in trouble can be very daunting to the decision-maker, who – despite the complexities and uncertainties – must still act. In this regard, Allen (2000: 10) argues that, in a radically immanent and complex world, the best one can hope to do is:

> to put in place the mechanisms that allow us always to question our "knowledge" and continue exploring. We must try to imagine possible futures, and carry on modifying our views about reality and about what we want.

M. Woermann, *On the (Im)Possibility of Business Ethics*, Issues in Business Ethics 37, DOI 10.1007/978-94-007-5131-6_4,

A position informed by a critical understanding of complexity, as well as Derrida's notion of the 'impossible' as that which possibilises notions such as ethics, responsibility, and justice, furthermore point to the need to develop cognitive strategies that are not limited by the logic of concepts. Morin (1974) refers to these cognitive strategies as 'the logic of complexity' (571), which he characterises as a style of thought that 'in its most essential moments... escapes the binary logic of all or nothing' (573). Morin is not blind to the logic of concepts (which, to recall, means that we cannot form concepts outside of this binary logic). However, he seeks to overcome the restrictions of a binary logic, as is clear from his appropriation of the following words by Nietzsche: 'the fact that we are unable to state and deny the same thing simultaneously in no way expresses a necessity but merely an inability' (in Morin 1974: 573–574). Morin (574) writes that the logic of complexity 'does not of course negate logic within the sphere in which it is operational, but... sublates it, in the Hegelian sense, that is retains it while integrating it in a richer logic.' As argued in Chap. 3, Derrida (1988) also supports this view of a broader logic, in stating that, despite the necessity of a binary logic, we can nevertheless add 'a supplementary complication', which:

> calls for other concepts, for other thoughts beyond the concept and another form of "general theory," or rather another discourse, another "logic" that accounts for the impossibility of concluding such a "general theory" (117).

Morin (1974) argues that the logic of complexity is dialectical, probabilistic and flexible. However, instead of elaborating on his ideas of how one is to conceive of this logic, we instead turn to what I have termed three important 'virtues' for a complex world, namely transgressivity, imagination, and irony (see Woermann and Cilliers, 2012: 453–459). These 'virtues' should not be viewed in terms of moral principles that should be habitually applied (as is the case in Aristotle's virtue ethics), but rather as mechanisms or attitudes that support the critical project and the meta-position that is being developed in this study. In other words, taking cognisance of these three mechanisms can help us to practically engage with the demands of a deconstructive and provisional account of ethics, and can also draw attention to the logic of complexity that serves to contaminate our efforts at concluding a general theory.

Transgressivity

The first mechanism that characterises a complex and critical position is that of transgressivity. Preiser and Cilliers (2010) write that a critical and provisional position can never simply re-enforce that which is current; and, in this regard, transgressivity serves as the means by which on can violate accepted or imposed boundaries. Transgressivity demands bold action, which – at a first glance – seems to be in contradiction with the modesty that is needed to accept the limitations of one's own normative and theoretical positions. However, one cannot practice transgressivity responsibly (i.e. commit to the ethical testimony implied in deconstruction) without an attitude of modesty. In other words, modesty and transgressivity go hand-in-hand, because modesty acts as the impetus for transgressivity, in focusing attention on the possibility of other rules of action.

Moreover, being transgressive is not only an ethical move, but also a political move. Both deconstruction and the provisional imperative place us under a moral obligation to stay open to the future and to the to-come, whilst *simultaneously* compelling us to respond to the urgency of the situation by taking in a position and assuming action. Derrida (2002) calls this the *aporia* of ethics and politics (which makes contradictory, but equally urgent, demands on the moral agent). The double-logic of ethics and politics marks the heart of the critical position, and transgressivity is situated in the fold of this *aporia*.

Although this *aporia* will be elaborated on in the following chapter at the hand of Derrida's own writings, it is nonentheless fruitful to recall Alain Badiou's (2009) becautiful tribute to Derrida, in order to further demonstrate the importance of transgressivity in Derrida's work. Badiou writes that:

> what is at stake in Derrida's work, in his never-ending work, in his writing, ramified as it is into so many varied works, into infinitely varied approaches, is the *inscription of the non-existent*. And the recognition, in the work of inscribing the non-existent, that its inscription is, strictly speaking, impossible. What is at stake in Derrida's writing – and here 'writing' designates a thought-act – is *the inscription of the impossibility of the non-existence as the form of its inscription* (132).

In his tribute, Badiou describes Derrida as the opposite of a hunter: whereas the hunter hopes that the animal will arrest its movements so that it can be shot, Derrida's animal cannot cease fleeing, since grasping the animal means suppressing it. This is because the very alterity that defines the animal as Other is destroyed the moment we attempt to assimilate it. Consequently, Badiou (133) writes that '[t]he vanishing point cannot be grasped *qua* vanishing point. It can only be located.' Deconstruction therefore operates in service of ethics, which can also be defined as the endless flight. Pursuing the endless flight and the Other – attempting to locate the Other – is a means of trying to account for, and thereby do justice to, the inexistent (which Badiou emphatically states is not the same as nothingness). The nature of the inexistent does not conform to a binary logic. Rather, the desire for inexistence is supported by what Badiou (141) calls, 'the monstration of the slippage', that occurs between saying that 'the non-existent is' (which fails to convey that it does not exist) and 'the non-existent does not exist' (which fails to convey that it is). In this example, we can clearly see the logic of complexity at play.

Badiou's tribute to Derrida lies in his decision to write inexistence with an 'a' (*inexistance*) in a similar fashion to *différance*, in order to attempt to 'couch non-existence' (143), and give it a place, which is also a non-place.

Returning to the *aporia* of ethics and politics, we can say that, on the one hand, *inexistance* implies ethics, because grasping fleeing as vanishing point entails an endless vigilance. On the other hand, however, one has a duty to actively transgress existing boundaries or limits and give voice to the non-existent, typified for Badiou (141) in the war cry of a Revolution, which rings 'We are nothing, let us be all!'

In other words, one has a duty to take a position and engage in politics. This example clearly illustrates how the transgressive nature of the critical position is fed by the slippages of ethics, but leads to the binaries of politics; and, similar to the way in which Derrida's limit concepts operate in the world, ethics and politics are also joined in a relation of 'unrevisable separation and permanent provocation' (Teubner 2001: 42). With regard to the nature of transgression, Derrida (1981: 12) states:

> There *is not* a transgression, if one understands by that a pure and simple landing into a beyond of metaphysics... Now, even in aggressions or transgressions, we are consorting with a code to which metaphysics is tied irreducibly, such that every transgressive gesture reencloses us – precisely by giving us a hold on the closure of metaphysics – within this closure. But, by means of the work done on one side and the other of the limit the field inside is modified, and a transgression is produced that consequently is nowhere present as a *fait accompli*. One is never installed within transgression, one never lives elsewhere. Transgression implies that the limit is always at work.

What should be clear from the above analysis is that transgressivity defines responsible and considered decisions in a complex and uncertain world. Although this point applies generally, the importance of transgressive actions in the workplace specifically should not be underestimated. In this regard, Olivier Babeau (2007) argues that deviance (which in this context functions as a synonym for transgressivity) is necessary both for the successful completion of work tasks, and for responding to the abuse of power. Although Babeau does not explicitly account for the dual ethical and political demands that mark transgressive actions, his analysis – as will be shown – is not incompatible with the position of trasnsgressivity provided above; and, moreover, give a practical illustration of the importance of transgressivity in our daily lives.

Transgressivity in an Organisational Context

With regard to his first point, Babeau (2007) argues that strict adherence to rules can serve as an inhibitory obstacle to the successful completion of work tasks, because certain tasks – especially complex tasks or tasks that unfold unpredictably – require a degree of deviance or transgressivity. To illustrate this point, Babeau cites the example of air traffic controllers. He writes that '[c]ontrollers flout the security norms in order to reconcile two contradictory needs, which are traffic flow and safety' (37). The infinite vigilance required in terms of securing passengers' safety is always interrupted by the urgent need for a response and for action, otherwise it is impossible to successfully enforce air traffic control. Air traffic controllers therefore often defy the establishment's rules, and act according to parallel, implicit, and informal norms, which are both enabling and transgressive. What this example shows is that formal rules and codes are often insufficient for dealing with the complex realities that characterise our working lives, and this point is also reinforced by Babeau's argument that transgression is a necessary counterpoint to formal and formalised organisational power structures (characterised by manifest ideologies and political positions).

With regard to this point, Babeau argues that transgression creates 'liberty spaces', characterised by 'fuzziness and ambiguity' (45), which serve to balance the relationship between organisational actors. In other words, organisational culture itself cannot be fully described in formal terms, since such accounts cannot account for the logic of complexity, which Morin (1974: 575) argues can help us to deal with sets 'in an uncertain and oscillatory manner'. Babeau (2007) further argues that these 'fuzzy sets', which are facilitated by transgression, serve to reconcile autonomy and liberty in organisations. In this regard, one could argue that transgressivity creates the space for ethical and responsible actions by opening up reified patterns of hierarchy and authority.

Babeau concludes in arguing that rules and deviance (or transgression) stand in a dialectical relationship, and should not be viewed as a binary construct. However, Babeau argues that *vague rules*, which Marcel Duc and Daniel Faïta (1996: 65; trans. O. Babeau) define as 'global rules given by the chief which contain the conditions of their own contradiction', serve to capture this dialectical relationship. In this regard, Babeau gives the example of a building site, where the supervisor issues the day's instructions, well-knowing that these instructions will be transgressed as contingencies arise on site. As such, the instructions serve as a vague or orientating rule, which rests on a type of performative contradiction. Although this example is not normative in nature, it supports the stance towards standard normative ethical theories, developed in this study. To recall: responsible action demands that we acknowledge, but also transgress, the rule or theory in the moment of the decision. It requires a degree of invention, and risk, and it requires that we adopt a critical attitude towards codes and rules.

More generally, Babeau's discussion on transgression is an organisational context also serves to reinforce the point that organisations are constituted by contradictory (as opposed to rational) practices and norms. In other words, organisations are complex institutions, and this complexity can only be successfully dealt with through acknowledging that organisational boundaries, and technical and ethical rules and norms, cannot be drawn in an unproblematic fashion. Transgressive action is necessary precisely because organisations and organisational life – like all other aspects of life – cannot be perfectly captured in any model.

Irony

Transgressivity is supported by an ironic outlook on life. Consider dictionary descriptions of irony as a means of expressing something other than the literal intention of words, as a demonstration of incongruence between what is expected and what is, and as an expression of that which is contrary to plan or expectation. In these descriptions, irony is defined as a means by which to subvert the idea of an objective reality. This is achieved by introducing an element of contingency and play into literal, objective language. The value of irony is that it draws attention to the supplementary complications that govern all rules (and hence to the logic of complexity) by showing how the meaning of our concepts is always-already contaminated by other meanings. As such, irony affirms the necessity of improvising when faced with the limits of a binary logic, and therefore also supports complexity thinking. Furthermore, in drawing attention to the fact that there is no final truth that allows us to objectively operationalise our actions, irony implies a self-critical position (Presier and Cilliers 2010).

Given this understanding, irony can be interpreted as a generative creative task, akin to a form of improvisation, and one can further argue that there is 'an important and potentially fruitful connection' between these skills and 'the lived experience of complexity' (Montuori 2003: 238). Alfonso Montuori (238) elaborates by stating 'that improvisation and creativity are capacities we would do well to develop in an increasingly unpredictable, complex, and at times chaotic experience.' It is specifically in relation to developing fruitful and responsible strategies for living that the virtue of irony is indispensible.

We are all improvisers who not only tell a story, but become a story. We create interwoven narratives, which, together, constitute a tapestry of stories (Montuori 2003; also see Kearney 1988). The ethical moment lies in whether we concede to this or not, i.e. whether we accept – with irony and humour – our limited knowledge and fragile personal experiences, and focus these in the very moment in which we are living (Montuori 2003). To be able to improvise and to live with irony:

> requires a different discipline, a different way of organizing our thoughts and actions. It requires, and at best elicits, a social virtuosity which reflects our state of mind, our perceptions of who we are, and a willingness to take risks, to let go of the safety of the ready-made, the already written, and to think, create, and 'write' on the spot (244).

Assuming an ironic disposition simultaneously draws attention to the status of our strategies, and lightens the burden of self-awareness. This is because those who live with irony find it easier to confess to the fact that their lives are not following a determinate course, but represent the outcome of choices and decisions. Irony – like transgressivity – needs modesty, which in this context means adopting a self-deprecating humour, and not taking oneself or one's ideas too seriously, as this may prevent one from exercising the openness needed to act responsibly in the face of complexity. Irony is a critical task, without which we potentially open the door to human evil. As Susan Sontag (2007: 227) suggests, it is exactly 'this refusal of an extended awareness' that lies at the heart of 'our ever-confused awareness of evil' and 'of the immense capacity of human beings to commit evil'. In a sense then, it is irony that allows us to face up to the seriousness of our responsibilities, which is an insight that accords beautifully with the description of irony as a demonstration of incongruity between what is expected and what is.

This analysis of irony supports our understanding of the double movement involved in both affirming a certain position, and denying the absolutist status of that which we are affirming; and, indeed, the use of irony in Derrida's own writing also reinforces this point. In this regard, Claire Colebrook (2004: 95) argues that one of the greatest achievements of Derrida's philosophy is its ability to create a link between two styles of irony: 'a satirical irony that attacks the conventions of a specific context, and a broader romantic or transcendental irony that aims to think beyond context.' As regards satirical irony, Colebrook argues that Derrida recognises the necessity and lawfulness of concepts (here, Derrida's argument regarding the necessity of binary constructs, language, authority, hierarchies, and law come to mind). Like Socrates, Derrida seriously engages with the text in order to push its limits and reveal its position to itself. Colebrook defines this activity as both ironic and counter-ironic, in that – unlike Socrates – Derrida employs our metaphysical and logocentric concepts in ways that render them impossible.

Colebrook however also argues that Derrida moves beyond this position in drawing attention to the way in which irony pervades all meaning, in that – far from being a special case of meaning that deviates from the core or literal meaning – all texts inherently contain the force to disrupt (or otherwise stated: no text can close in upon itself). All meaning is potentially ironic, in that every (con)text contains more than is understood at any given time. In other words, the complexity that defines (con)text generates, what Colebrook (107) refers to as, 'forces of differentiation and implication well beyond the speaking subject', which again affirms the point that all meaning is contingent and provisional. This view of irony, which constitutes a deconstruction of the opposition between literal and ironic language, opens up our understanding of our contexts in a way that allows us to transgress these very contexts.

Irony in an Organisational Context

Although a sense of humour (or wit) is sometimes listed as a business virtue (see Solomon 1992, 1999), the value of an ironic disposition in business has not received attention. In philosophy however the theme of irony is not new. As mentioned above, one of the earliest, and most well-known expressions of irony in philosophy concerns Socratic irony. Furthermore, the German philosopher, Friedrich Schlegel, popularised the notion of romantic or philosophical irony, which he employed as a 'more complex philosophical tool', used to shed light on the divided self and a multiplicity of perspectives that could potentially unlock the truth of the whole (Williams 2003). The concept of irony is also prevalent in the works of Hegel, Kierkegaard, Nietzsche, and – more recently – Richard Rorty, the latter arguing that 'irony is the only possible ethic of modern liberalism' (Colebrook 2004: 151). Despite all these well-known accounts of irony in the extant literature, we turn to a figure that is not generally recognised as an ironist in the philosophical literature, namely Claude Levi-Strauss's *bricoleur*, in order to illustrate the value of irony in an organisational context. The specific reason for highlighting the *bricoleur* as an ironic figure, is because of the forces of differentiation and implication that defines the relation between the bricoleur, and its purported 'opposite', the engineer.

A *bricoleur* is defined as a fiddler or tinker, and, by extension, as someone who makes creative and resourceful use of whatever materials are at hand. Levi-Straus (1966) distinguishes the *bricoleur* from the engineer, arguing that the former approximates the savage mind that operates in a closed universe and is therefore forced to do with the means at hand, whereas the latter approximates the scientific mind that operates in an open universe because he can develop new tools, and thereby construct the totality of the universe. In 'Structure, sign, and play', Derrida (1978) deconstructs the distinction between the engineer and the *bricoleur*, by arguing that because there is no absolute origin (i.e. no objective reality that we can access through our practices), the engineer remains a theological idea; or, more radically, 'the odds are that the engineer is a myth produced by the *bricoleur*' (285). Derrida's point is that the world is a complex place, and pretending that it is otherwise is also merely a *bricolage* or strategy for dealing with the world. As such, Derrida (285) writes:

> as soon as we admit that every finite discourse is bound by a certain *bricolage* and that the engineer and the scientist are also species of *bricoleurs*, then the very idea of *bricolage* is menaced and the difference in which it took on its meaning breaks down.

Through means of the above deconstruction, Derrida shows that the relationship between the bricoleur and engineer is not oppositional, but differential. It is because of this insight that the ironic figure can be modelled on the figure of the self-conscious *bricoleur*, who would be defined as someone who is capable of channeling contingency and difference in a productive and playful manner, and of creatively reconciling diverse strategies. Successful leaders are *bricoleurs*, to the extent that they are able to act as 'intellectual MacGyvers' (Croak 2011). Such leaders are capable of tapping into organisational and societal narratives and sub-cultures, in order to facilitate the development of 'a meaning-producing pastiche' (Croak 2011), which constitutes a sense of dynamic organisational identity. As such, leaders can employ irony in nurturing a sense of working community, which in itself is a deeply ethical activity, if ethics is understood as one of the ways in which we constitute ourselves and our practices.

Imagination

Derrida's discussion of imagination is similar to the above discussion of irony in the sense that his critique of the mimetic imagination (which seeks to mirror or imitate reality) constitutes a deconstruction of the opposition between reality and imagination, which is premised on the supposition that there is some 'essence' which distinguishes imagination from a reality existing either before or beyond it (Kearney 1988). In this deconstruction, Derrida dismantles the notion of the origin of meaning, and replaces it with a view of the world as a 'never-beginning, never-ending text' (290). This holds the following implications for our understanding of imagination:

> If there is thus no thematic unity or overall meaning to reappropriate beyond the textual instances, no total message located in some imaginary order, intentionality, or lived experience, then the text is no longer the expression or representation (felicitous or otherwise) of any *truth* that would come to diffract or assemble itself in the polysemy of literature (Derrida 2004: 268–269).

Imagination, like irony, is therefore repositioned in the flow of the ordinary. Although Derrida does not offer a productive reading of what such a notion of imagination might entail, one could argue that a view of imagination that draws attention to the status of, and attempts to creatively engage with, this 'never-beginning, never-ending text', would be consistent with Derrida's deconstructive position. Specifically, an account of moral imagination as a critical meaning-producing process that takes place within specific contexts defined by power and politics accords well with Derrida's broader project. Such an account is offered below at the hand of Patricia Werhane's and Timothy Hargrave's analyses of the nature and role of moral imagination in organisational contexts.

Imagination in an Organisational Context

Todd May (1995: 145) notes that, '[t]he terms in which one thinks of oneself and one's possibilities, the practical parameters of those possibilities, and the ease or difficulty in realizing them, are all social as well as individual matters'. Therefore, imagination is both an individual and an organisational or systemic activity, and both these dimensions of imagination come to the fore in Werhane's (1999) analysis of the role of the imagination in an organisational setting. Werhane starts by making a distinction between moral judgement and moral imagination, and argues that, on a psychological level, both are needed to actively and consciously engage in our practices. This is because, without moral imagination, we are unable to disengage from our contexts and creatively reflect on what is possible, yet without moral judgement and reasoning we are in danger of slipping into moral fantasy. Although the distinction between moral judgement and imagination can be viewed in terms of a rigid binary, a more productive interpretation (and one which is better aligned with the goals of this study) is to view jugdgement and imagination as simultaneous, and equally necessary, processes for actively and productively engaging in the text. This interpretation is consistent with Werhane's project, especially given the fact that she considers psychological imagination as a means of prompting self-reflection, and creatively evaluating new possibilities in terms of existing frameworks. Together, these functions allude to the deconstructive gesture, which allows us to 'transform the frameworks we apply when apprehending the world' (Cilliers 2005: 264).

On the organisational or systemic level, Werhane (2002) argues that moral imagination can help us to transgress the boundaries of the models that we develop of our social practices. This is achieved by challenging the traditions and practices of an entrenched system; and, through this process, working towards a new mental model of the given system. According to Werhane, such a process is social in nature, and requires proactive leadership, an understanding of various stakeholder perspectives, a focus on the network of relations that constitutes the system, and an evaluative perspective.

Hargrave (2009: 87) also develops a social understanding of moral imagination, which he defines as something that 'emerges through dialectical processes that are influenced by actors' relative power and political skill'. Moral imagination is also a skill that needs to be fostered and exercised within 'pluralistic processes in which multiple actors with opposing moral viewpoints interact, and [where] no single actor is in control' (90). An element of conflict is always present in imaginative activities because of the 'lived tensions between contradictory perspectives' (91). Although Hargrave views moral imagination in terms of a collective action model, his analysis of moral imagination also has implications for individuals. In this regard, Hargrave (91) argues that 'morally imaginative actors recognise and integrate contradictory moral viewpoints, and also integrate moral sensitivity… [of] contextual considerations.' Since these characteristics are also hallmarks of critical thought, one can argue that moral imagination is itself a critical activity (Woermann 2010).

Another characteristic of moral imagination is that it involves an element of uncertainty or risk. Far from being a form of creative abandonment, moral imagination necessitates that we critically project and plan for the future (Woermann, 2010). However, since this future cannot be known, and since uncertainty involves a real property of situations, we have to respond with judgement (Luntley 2003) and be open to, and tolerant of, others' opinions.

The notion of tolerance is often interpreted in a negative light in the extant literature, and is increasingly viewed in terms of a passive acceptance of differences or an unwillingness to engage in other perspectives (see Forst 2007). Indeed, Derrida (see Derrida and Ferraris 2001) is very critical of tolerance, and Raffoul (2008: 288) paraphrases his sentiments on the matter as follows:

> Tolerance, for instance, i.e., hospitality *up to a point*, is no hospitality, is in fact the "contrary" of hospitality: the other is here "welcomed" on the basis of the conditions laid out by the host, that is, by a welcoming *power*.

James Mensch (2003) however has a different take on the term, and explains that, in Latin, tolerance has the sense of supporting or sustaining, rather than enduring or suffering. He further states that tolerance 'can be understood as the attitude that actively sustains the maximum number of compatible possibilities of being human' (142). This definition is supported by Morin's (1999: 54) description of tolerance as something that 'implies that we have convictions and faith, make ethical choices, and at the same time accept the rights of others to express different or even opposite choices, convictions, and ideas.' Although this understanding of tolerance may not be radical enough to support the type of open-ended hospitality that Derrida has in mind, it certainly does seem to be an important virtue for sustaining morally-imaginative processes in working contexts. This is especially true to the extent that tolerance promotes openness to diversity and an acceptance of complexity – both of which are important in constructively engaging with the conflicts and dissensus that characterise organisational life.

Concluding Reflections

The above brief analysis of 'virtues' for a complex world is by no means exhaustive, but it does serve as an indication of how a complex, deconstructive notion of ethics necessitates that we supplement our rational tools of analysis with other tools, which are better capable of accounting for the logic of complexity. Indeed, if we cannot keep the descriptive (i.e. objective) and the normative dimensions apart (as has been argued in this study), then the traditional understanding of a theoretical 'tool' – as something that interfaces between the cognitive domain and objective reality – is subject to a deconstruction of sorts, in that the term is opened up to include not only analytic tools, but any strategy or operation that can help us to navigate our way through the complexities of our lived experiences. It is in light of this understanding, that the above three 'virtues' for a complex world is presented. Moreover, the brief explanation that was provided of the functioning and role of these 'virtues' in an organisational context, should be read as a general introduction or preface to how the theoretical position developed in this study can be operationaised in the context of the workplace. However, for a more careful and detailed argument, we turn to part II of the study, in which the notion of corporate social responsibility is re-evaluated in light of this theoretical basis.

References

Allen, P. 2000. Knowledge, ignorance, and learning. *Emergence* 2(4): 78–103.

Babeau, O. 2007. Granting disorder a place in ethics: Organization's deviant practices and ethics. In *Business ethics as practice: Representation, reflexivity and performance*, ed. C. Carter, S. Clegg, M. Kornbergger, S. Laske, and M. Messner, 32–48. Cheltenham: Edward Elgar Publishing.

Badiou, A. 2009. *Pocket Pantheon*, trans. D. Macey. London: Verso.

Cilliers, P. 2005. Complexity, deconstruction and relativism. *Theory, Culture & Society* 22(5): 255–267.

Colebrook, C. 2004. *Irony: The new critical idiom*. London: Routledge.

Croak, J. 2011. Briocoleur. *Edge World Question Center*. Available online at: http://www.edge.org/q2011/q11_8.html.

Derrida, J. 1978. Structure, sign, and play in the discourse of the human sciences. In *Writing and difference*, trans. A. Bass, 278–294. Chicago: University of Chicago Press.

Derrida, J. 1981. Positions. In *Positions*, trans. A. Bass, 37–96. Chicago: University of Chicago Press.

Derrida, J. 1988. Afterword. In *Limited Inc.*, ed. G. Graff and trans. S. Weber, 111–160. Evanston: Northern Western University Press.

Derrida, J. 2002. The right to philosophy from a cosmopolitical point of view: The example of an international institution. In *Ethics, institutions, and the right to philosophy,* ed. and trans. P.P. Trifonas, 1–18. Maryland: Rowman & Littlefield Publishers.
Derrida, J. 2004. Plato's pharmacy. In *Dissemination,* ed. and trans. B. Johnson, 67–186. London/ New York: Continuum.
Derrida, J. and Ferraris, M. 2001. *A taste for the secret,* ed. G. Donis and D. Webb. Malden: Polity Press.
Duc, M. and Faïta, D. 1996. Savoir-faire d'encadrement et prescription floue. In *La transgression des règles au travail,* ed. J. Girin, 51–81. Paris: l'Harmattan.
Forst, R. 2007. '"To tolerate means to insult": Toleration, recognition, and emancipation.' In *Recognition and power*, ed. B. Van den Brink and D. Owen, 215–237. Cambridge: Cambridge University Press.
Hargrave, T.J. 2009. Moral imagination, collective action, and the achievement of moral outcomes. *Business Ethics Quarterly* 19(1): 87–104.
Kearney, R. 1988. *The wake of imagination: Towards a postmodern culture.* Minneapolis: University of Minnesota Press.
Levi-Strauss, C. 1966. *The savage mind*, trans. J. Weightman and D. Weightman. Chicago: University of Chicago Press.
Luntley, M. 2003. Ethics in the face of uncertainty: Judgement not rules. *Business Ethics: A European Review* 12(4): 325–333.
May, T. 1995. *The moral theory of poststructuralism.* Pennsylvania: Pennsylvania State University Press.
Mensch, J.R. 2003. *Ethics and selfhood.* New York: State University of New York Press.
Montuori, A. 2003. The complexity of improvisation and the improvisation of complexity: Social science, art and creativity. *Human Relation* 56(2): 237–255.
Morin, E. 1974. Complexity. *International Social Science Journal* XXVI(4): 555–582.
Morin, E. 1999. *Seven complex lessons in education for the future*, trans. N. Poller. Paris: UNESCO Publication.
Preiser, R., and P. Cilliers. 2010. Unpacking the ethics of complexity: Concluding reflections. In *Complexity, difference and identity*, ed. P. Cilliers and R. Preiser, 265–287. Dordrecht: Springer.
Raffoul, F. 2008. Derrida and the ethics of the impossible. *Research in Phenomenology* 38: 270–290.
Solomon, R.C. 1992. *Ethics and excellence: Cooperation and integrity in business.* New York: Oxford University Press.
Solomon, R.C. 1999. *A better way to think about business.* Oxford: Oxford University Press.
Sontag, S. 2007. At the same time: The novelist and moral reasoning: The Nadine Gordimer lecture. In *At the same time: Essays and speeches,* ed. P. Dilonardo and A. Jump, 210–230. New York: Farrar, Straus and Giroux.
Teubner, G. 2001. Economics of gift – Positivity of justice: The mutual paranoia of Jacques Derrida and Niklas Luhmann. *Theory, Culture & Society* 18(1): 29–47.
Werhane, P.H. 1999. *Moral imagination and management decision-making.* New York: Oxford University Press.
Werhane, P.H. 2002. Moral imagination and systems thinking. *Journal of Business Ethics* 38(1/2): 33–42.
Williams, Z. 2003. The final irony. *The Guardian,* Saturday 28 June 2003.
Woermann, M. 2010. Corporate identity, responsibility and the ethics of complexity. In *Complexity, difference and identity*, ed. P. Cilliers and R. Preiser, 167–192. Dordrecht: Springer.
Woermann, M. and Cilliers, P. 2012. The ethics of complexity and the complexity of ethics. *South African Journal of Philosophy*, 31(2): 447–463.

Part II
Practical Application

Chapter 5
Reconsidering the Meaning of Corporate Social Responsibility

Abstract In this chapter, the implications that a deconstructive and complex ethics hold for our understanding of corporate social responsibility (CSR) will be investigated. The chapter commences with an overview of the traditional characterisation of CSR, in order to demonstrate that this characterisation is informed by an equalising or commutative understanding of justice, i.e. repaying good with good. On this interpretation, CSR policies articulate the content of the social contract, which is premised on a commutative understanding of just relations between societal and economic interest groups. Derrida however offers a much more radical view of responsibility, one that transcends the reciprocal demands and expectations of a circular economy. Responsibility, on his take, becomes an expression of ethical complexity, which means that, in practice, responsible action always pushes the limits of its own expression. However, this understanding of responsibility cannot form the basis of a substantive ethics, and is often criticised for being practically useless. More specifically, critics are concerned that if a Derridean view of ethical relations and responsible action are irreducible, undecidable, and non-subsumptive, then it is not clear on what basis moral judgement can take place (the charge of relativism), or of what value business ethics can be (the charge of irreducibility). Both these charges are addressed in this chapter at the hand of a close reading of Derrida's work, in order to show how these charges can be overcome, but also to illustrate the value that a complex deconstructive ethics holds for business ethics in general and CSR in particular.

Introduction

In Part I of this study, the theoretical basis for a deconstructive, complex understanding of business ethics was developed. In Part II (beginning with this chapter), this basis is applied to prominent business ethics themes, in an attempt

M. Woermann, *On the (Im)Possibility of Business Ethics*, Issues in Business Ethics 37,
DOI 10.1007/978-94-007-5131-6_5,

to operationalise the insights gleaned thus far, and, where necessary, to extrapolate on these insights. It is specifically the idea of corporate social responsibility (CSR) that will receive attention, since our view of CSR informs our understanding of related themes such as corporate governance and accountability, stakeholder theory, corporate identity, and responsible leadership. Prominent views on CSR differ, and as such we cannot postulate a universal theory of CSR with which to compare a complex, deconstructive account of CSR. In order to proceed with some form of comparative analysis, the strategy that is followed in this chapter is to try and determine the central tenets of popular CSR accounts, in order to highlight how a deconstructive, complex view of CSR differs from these accounts. The criticisms routinely lodged against a deconstructive account of business ethics will be addressed, and the value in taking on such a view of business ethics will be highlighted. After having made the case for the merits of a deconstructive, complex ethics, a theory and model for CSR will be proposed in the following chapter.

Before embarking on the analysis it is important to note that, despite the large amount of literature on CSR, the concept is not easily definable. Dirk Matten and Jeremy Moon (2008) offer three prominent reasons for this: firstly, CSR is an evaluative or valued concept, which is internally complex, with relatively open rules of application; secondly, CSR is an umbrella terms, which overlaps, or is used synonymously, with other concepts that articulate the relation between society and business; and, thirdly, CSR is a dynamic concept, the content of which is contingent on contextual and historical factors. Nonetheless, Matten and Moon (405) offer the following broad description of CSR:

> At the core of CSR is the idea that it reflects the social imperatives and the social consequences of business success. Thus, CSR (and its synonyms) empirically consists of clearly articulated and communicated policies and practices of corporations that reflect business responsibility for some of the wider good. Yet, the precise manifestation and direction of the responsibility lies at the discretion of the corporation. CSR is therefore differentiated from business fulfilment of core profit-making responsibility and from the social responsibilities of government (Friedman 1970).

In this description, CSR is identified as a concept that sits between the economic and state spheres, and that implies a normative dimension (in terms of expressing businesses' responsibility for certain social imperatives and for the social consequences that arise from business success), an operational dimension (in terms of articulating and evaluating business responsibilities), and discretionary responsibility (in terms of giving specific content to, and acting on, social imperatives). In this analysis, the focus will be on giving a deconstructive reading of the scope and nature of corporate responsibility, and the concurrent degree of discretionary responsibility that this notion of corporate responsibility implies. (The operational dimension of CSR will be addressed in the next chapter.) Before turning to this account of CSR, it is firstly necessary to provide an overview of traditional characterisations of CSR.

Characterising Traditional Accounts of CSR

In his article, 'How corporate social responsibility is defined: An analysis of 37 definitions', Alexander Dahlsrud (2006) identifies the five main dimensions of CSR through undertaking a content analysis of existing CSR dimensions. These dimensions are: the *environmental dimension*, which pertains to definitions of CSR that refer to the natural environment; the *social dimension*, which refers to definitions that articulate the relationship between business and society; the *economic dimension*, which comes prominently to the fore in definitions of CSR that highlight socio-economic or economic aspects of business responsibility, including describing CSR in terms of core business; the *stakeholder dimension*, which characterises CSR definitions that focus mainly on expressing stakeholder interests; and, the *voluntariness dimension*, which pertains to definitions that prescribe corporate action that exceeds the requirements of law (such as for example acts of corporate philanthropy). The relative importance of these dimensions is contingent on the view of CSR that one subscribes to, and to illustrate this point, a number of distinctions will be drawn, in order to demonstrate the dominant and competing views of CSR.

Narrow vs. Broad Views of CSR

A first prominent distinction in the extant literature concerns the scope and proper domain of CSR. The two rival views in this regard are the narrow view of profit maximisation, and the broad view that accounts not only for shareholder or stockholder interests, but also for business and societal relations. Hence, these two views express the *economic* and *social dimensions* of CSR respectively.

The societal case for CSR is articulated in what is commonly recognised as the first attempt to theorise corporate responsibility, namely Howard Bowen's 1953 publication, entitled *Social Responsibilities of the Businessman*. Ming-Dong Paul Lee (2008) notes that, in this book, Bowen argues that because of the great influence and far-reaching scope of business decisions and actions, businesses are obligated to consider the social imperatives and consequences of business success. Of interest for Bowen, is the exact nature of these responsibilities, and how business could go about making the necessary institutional changes to promote CSR. Lee also notes that this publication corresponded with the New Jersey Supreme Court's landmark ruling, in which corporate contributions beyond the profit motive were legitimised.

In the two decades following Bowen's seminal publication, many theorists sought to address his questions concerning the content and processes of CSR. However, many also challenged his core assumptions regarding the scope of businesses responsibilities, arguing against the *prima facie* acceptance of corporate responsibility for societal interests (Lee 2008). The narrow view of CSR is most forcefully and famously articulated by the economist Milton Friedman, in his publication entitled *Capitalism and Freedom* (1962). Herein, he argues that:

> The view that has been gaining widespread acceptance [is] that corporate officials... have a social responsibility that goes beyond serving the interest of their stockholders... This view shows a fundamental misconception of the character and nature of a free economy. In such an economy, there is one and only one social responsibility of business – to use its resources and engage in activities designed to increase its profits as long as it stays within the rules of the game, which is to say, engage in open and free competition, without deception or fraud... Few trends could so thoroughly undermine the very foundations of our free society as the acceptance of corporate officials of a social responsibility other than to make as much money for their stockholders as possible (133).

Friedman's reasons for arguing that CSR will undermine the foundations of our free society are fourfold: firstly, the acceptance of corporate responsibility by executives undermines the promissory relation, which according to Friedman exists between executives and stockholders, and which amounts to the agreement that executives will act in the economic interests of stockholders. Secondly, the acceptance of CSR obligations by corporate executives undermines the function of business in that it hampers the effective functioning of the market mechanism. Thirdly, exercising CSR in effect turns business into an arm of the state (since business is now fulfilling government's duties). This is unethical, because executives impose a 'tax' on stockholders, in using their money for social objectives. Lastly, unlike government officials, executives are not democratically-elected, and therefore do not have the right to make decisions regarding the allocation of funds for social objectives.

Although Friedman's argument is still considered significant, Lee (2008) argues that modern corporate equity holding patterns have resulted in a shift in the meaning of stockholder interests, typified by the idea of 'enlightened self-interest' that characterised thinking on CSR in the 1970s. According to this view, CSR is considered consistent with stockholders' long-term interests, and since corporate equity holding patterns have diversified, it is in the interest of stockholders to spread social expenditure across firms (see Wallich and McGowan 1970). However, a second reason that undermines the validity of Friedman's argument concerns the nature of the social contract.

The social contract expresses the implicit relation between business and society, whereby society grants business the 'license to operate' through public consent, in the expectation that business – defined as 'productive organizations' – will address certain societal needs (including the satisfaction of consumer and worker interests) (Donaldson 1993). The exact content of the social contract is however contingent on the specific context in which businesses operate. Whereas in Friedman's day, economic returns and sound business practices might have constituted the main terms of the social contract, Melvin Anschen (1970: 10) noted around this period that soon 'it will no longer be acceptable for corporations to manage their affairs solely in terms of the traditional internal cost of doing business, while thrusting external costs on the public.'

Indeed, today the social contract reflects our current concerns regarding quality of life and environmental sustainability (Shaw 2008), and, it is – at the very least – expected that companies will fulfil their negative duties, in terms of accounting for the negative externalities that arise from business operations. In other words, the *environmental dimension* of CSR policies is also gaining increasing significance. More often than not, society also expects business to contribute positively to the welfare of society (*the social dimension*), either through their core business or through philanthropic activities. This broader approach to CSR is consistent with the theoretical position espoused in this study because, although stockholder interests remain important, it is also imperative that corporations take cognisance of the larger operating environment in which they are embedded, and which they affect through their actions.

Normative vs. Instrumental Justifications for CSR

A second distinction that can be drawn concerns the normative versus instrumental justifications for CSR. Whereas Bowen's justification for the acceptance of corporate responsibility is distinctively normative in nature, those who follow a view of 'enlightened self-interest', take on CSR obligations because they believe that these obligations improve business success. One of the best contemporary expressions of normative CSR is Edward Freeman's work on stakeholder theory. The basic premise of stakeholder theory is that the survival of the corporation trumps both social and economic goals, and the best way in which to ensure the future survival of the corporation is through paying heed to the *stakeholder dimension,* which means that organisations should account for how the corporation affects not only stockholders, but also stakeholders (such as employees, suppliers, customers, government etc.). Lee (2008) argues that stakeholder theory has moved to the centre stage of research in business and society relations. Although stakeholder theory need not be justified on normative grounds, Freeman (2008) argues in favour of the normative approach on the basis of what he terms the 'Open Question Argument' (43).

Freeman (2008) states that many accounts of CSR, and indeed of business ethics, work with the assumption that business and ethics are separate from each other. He refers to this view as the separation thesis, arguing that this thesis rests on the fallacy that business decisions bear no ethical content, and vice-versa. Although still popular in many accounts of business, Freeman rejects this view in favour of the 'Open Question Argument'. This argument serves to encourage business practitioners to consider questions regarding which stakeholder groups will be affected by their business decisions, as well as how their business decisions affect them and these stakeholder groups. The 'Open Question Argument' is based on the integration thesis, wherein it is accepted that most business decisions also have ethical content, and that ethics involves accepting responsibility for how our decisions affect the lives of others. This is also consistent with the view of ethics developed in this study, since the integration thesis recognises the normative dimensions of all decisions – business or otherwise. Whereas the separation thesis makes possible instrumental justifications for business ethics (good CSR practices are good for business, and hence desirable), the integration thesis is supported by normative justifications for business ethics (good CSR practices are constitutive of good business, and hence necessary).

In instrumental accounts, CSR is viewed as a strategic resource that can improve corporate performance. Lee (2008) however argues that there has been a shift in the way that corporate performance is understood: whereas bottom-line considerations used to dominate views on corporation performance, social (and even environmental) considerations are playing an increasingly important role (in this regard, see John Elkington's (1999) work on triple-bottom line reporting). Lee also notes three shortcomings of the instrumental or business case for CSR: Firstly, after 30 years of research there is still no definite causal link between CSR and profit, and it is doubtful whether future research will be able to prove this link (especially since, as more and more companies become socially responsible, the marginal value of CSR practices decreases). Secondly, instrumental reasons alone cannot account for organisational changes regarding CSR. Too little attention is being paid to the personal ethics of managers, the influence that institutional changes exert on corporations, and the impact of the socially responsible investment movement on corporate practices. Thirdly, the instrumental view uncritically accepts that what is good for business must also be good for society, and may lead to biases with regard to how companies select CSR objectives. In other words, 'bias will result in increased corporate attention to certain social needs that are less costly and potentially profitable, while other more costly social misery [sic] will be conveniently ignored' (Lee 2008: 65). That is, the instrumental case for CSR risks pushing the *economic dimension* of CSR at the expense of the *social dimension*, and of excluding the complexities that define good CSR practices.

CSR as Voluntarism vs. CSR as Core Business

Not only are there different views regarding the scope of, and justification for, CSR, but differences also prevail with regard to how CSR should be operationalised, especially when CSR obligations are extended beyond the profit motive. Since first being coined in 1953, the concept of CSR has gradually converged with corporate performance; and, as Lee (2008: 63) notes, this trend has worked in both directions:

> On the one hand, the concept of CSR has expanded to envelop both economic and social interests on macro-political as well as organizational levels. On the other hand, the concept of corporate performance also broadened to cover economic as well as social interests on institutional as well as organizational levels.

The effects of this trend are particularly evident in terms of the content of CSR practices. Whereas in the past, corporate philanthropy was most often associated with good CSR practices (the *voluntariness dimension*), today core business or competencies are increasingly being harnessed for developmental impact (the *economic* and *social dimensions*). This shift signifies a greater integration of CSR with business practices. In the voluntarism-model, core business continues to deliver stockholder value, whereas philanthropy or isolated CSR-practices deliver stakeholder value; in the core-business model, companies seek to deliver both stockholder and stakeholder value through corporate activities (Ashley 2009).

Caroline Ashley (2) notes that initiatives to adapt core business for developmental impact have only started developing in the last 10 years, and include concepts such as 'Bottom of the Pyramid' (BOP) (where companies market and develop products for billions of poor consumers, in order to make profits whilst reducing poverty); 'Creative Capitalism' (which was coined by Bill Gates, and which refers to capitalism that works by generating profits whilst solving social inequalities); 'Social Business' (in which profitable businesses reinvest into companies that seek to help the poor); 'Ethical Trade' (which is concerned with decent labour standards); and, 'Inclusive Business' (which has less ethical connotations than CSR and embraces wider concepts, such as marketing for the poor).

Ashley argues that the core business approach (also referred to as 'inclusive business') extends traditional CSR practices, but also excludes certain traditional CSR elements, and therefore overlaps, but is not synonymous with, what is generally considered to be CSR. The main argument for the core business approach is that, unlike certain CSR practices, the acceptance of wider responsibilities are not viewed as an add-on to business, but integral to business. Also, although traditional CSR practices are often only associated with the fulfilment of negative duties, the core business approach moves beyond problem mitigation to embrace and develop opportunities for the previously excluded. In practice however, Ashley argues that the clear distinction between CSR and the core business approach is problematised, in that this approach often incorporates standards of responsible business practices, which are key to CSR; the core business approach is viewed as one type of CSR practice; and, to initiate projects related to core business often requires the expertise of staff in the CSR department, who have experience with developmental projects beyond mere operational deliverables.

In the future, we are likely to see an increasing emphasis on the core business approach as a means for companies to meet their CSR obligations. The core business approach also subscribes to the integration thesis, in that the normative consequences of business practices are considered and developed. However, CSR is no longer framed in the language of ethics, and in this regard, one should bear in mind Lee's (2008) warning that – in recasting CSR as core business – we risk focusing our attention on those aspects of development that are profitable for the company, possibly at the expense of more pressing social and developmental issues. At best, the core business approach represents a win-win solution for business and society; at worst, it runs the risk of subverting normative considerations to economic objectives.

Explicit-Individualist vs. Implicit-Collectivist Accounts of CSR

The various perspectives on CSR discussed above should also be viewed within the historical development of CSR. Therefore, a last prominent distinction concerns the differences between the two broad accounts of CSR that have developed in response to the different contextual factors at play in the American and European business landscape.[1] De George (2008) argues that, despite broad areas of overlap in North American and European societies, corporations are generally more integrated into the social fabric in Europe, than is the case in North America. As a result, North Americans follow a more individualistic approach to business, whereas Europeans are more collectively-orientated when it comes to business. This has a profound effect on the way in which both business ethics and CSR are viewed.

With regard to business ethics in general, De George argues that, in the USA, ethical theories, in which precedence is given to the individual moral agent, are commonly used to evaluate new and existing business practices. De Georges lists the theories as comprising the work of Kant, Mill, Aristotle, Rawls, the pragmatists, the feminists, and theories of rights and justice. In terms of European business ethics, the emphasis is on structure and systems, and employee-codetermination. Such an approach is well-captured in the theory of communicative action espoused by Habermas; but, depending on how it is conceptualised, may also be encapsulated in theories that give expression to the complex ways in which business and society are related.

In terms of CSR specifically, De George (2008: 76) argues that 'in trying to gain a clearer perspective on CSR, we [must] start by emphasizing that corporations are constructions of society' and that '[u]nless they are seen as such, it is difficult to justify society's placing requirements on them.' Due to the different social structures and systems at play in Europe and the United States, CSR also took on different forms in these societies. For example, environmental issues constitute a central concern in Europe, but not in North America; and the issue of discrimination is differently framed in these two contexts, because Europe does not have the same history of slavery as does the United States. De George (76) raises these examples to make that point that:

> corporate social responsibilities, to the extent that they are not ethical or moral responsibilities, reflect the expectations and demands of the societies in which the corporations are found and/or where they operate.

Matten and Moon (2008) have also investigated differences in conceptions of CSR on each side of the Atlantic, specifically differences in articulating CSR commitments. These differences are denoted by their terms 'explicit CSR' and

---

[1] As the notion takes on global significance, we can expect to see more contextually-defined accounts of CSR, although the significance of such accounts may be limited by the current hegemonic power of the West.

'implicit CSR'. Explicit CSR refers 'to corporate policies that assume and articulate responsibility for some societal interests', whereas implicit CSR refers 'to corporations' role within the wider *formal and informal institutions* for society's interests and concerns' (409), and is therefore more compatible with the position espoused in this study. Whereas explicit CSR normally refers to voluntary programmes, policies, and strategies that articulate the relation between society and business; implicit CSR normally consists of values, norms, and rules that guide corporations' treatment of stakeholders, and that help to define corporate obligations in a collective sense. Explicit CSR frames CSR issues in prescriptive language, and constitutes deliberate, voluntary, and often strategic expressions of CSR commitments; whereas implicit CSR constitutes a reflective or reactive engagement with the institutional environment. Matten and Moon further link explicit and implicit CSR to the individualist and collective culture of the United States and Europe respectively. In this regard they argue that '[i]nstitutions encouraging individualism and providing discretion to private economic actors in liberal markets would be considered national systems in which one would expect to find strong elements of explicit CSR' (410). Conversely, institutions that have 'co-ordinated approaches to economic and social governance through a partnership of representative social and economic actors led by government' (410) contain elements of implicit CSR.

Despite these contextual differences, Matten and Moon note that explicit CSR is also beginning to gain traction in Europe and in the rest of the world. In order to explain this trend, they employ neo-institutional theory, which is based on the argument that 'organizational practices change and become institutionalized because they are considered legitimate' (411). Firstly – according to neo-institutional theory – rules, norms, and laws (referred to as 'coercive isomorphisms') assign legitimacy to new practices, and in the case of CSR, self-regulatory and voluntary mechanisms have proliferated in Europe in recent times. Secondly, given the uncertainties and complexities of the business environment, managers view 'best practices' in their organisational field (referred to as 'mimetic processes') as legitimate. As a result, we see an explosion of CSR reports in Europe, a large European representation in the U.N. Global Compact initiative, an increase in CSR training programmes, and a number of business coalitions for CSR. Thirdly, educational and professional authorities (referred to as 'normative pressures') also serve to legitimise organisational practices. In this regard, pertinent examples are the inclusion of a CSR component in most MBA programmes, a growing number of European professional associations, and the establishment of the European Academy of Business and Society. Matten and Moon (412) conclude this discussion, arguing that:

> Shifts in the balance of implicit and explicit CSR therefore reflect changing features of corporations' historical national institutional frameworks and their immediate organizational fields... The corporation is both embedded in its historically grown national institutional framework... as well as in its organizational field, which influences the corporation through isomorphic forces. The result is CSR reflecting a hybrid of implicit and explicit elements.

What the discussion shows is that the normative dimensions of CSR cannot be separated from the descriptive (i.e. the institutional, organisational, and historical) dimensions, which also shape our views on, and practices of, CSR. In this regard, Dahlsrud (2006: 6) argues that the challenge for business does not lie in defining CSR, but in understanding 'how CSR is socially constructed in a specific context and how to take this into account when business strategies are developed.'

Derridean Responsibility: The Deconstructive Challenge to Traditional Conceptions of CSR

Reconsidering the Meaning of Ethics and Responsibility

Given the above overview of traditional conceptions of CSR, it is clear that some of the main questions that dominate the debate pertain to the scope of corporate responsibilities, the justifications for assuming responsibility, how corporate responsibilities should be enacted, and the different social imperatives that shape corporate responsibilities. We now return to De George's question, posed in the concluding section of the previous chapter, which, to paraphrase, is: how do the categories made available in Derrida's work (particularly the complex interpretation given of his work in this study) extend our understanding of CSR?

A clue to answering this question is found in De George's (2008) own article, entitled 'An American perspective on corporate social responsibility and the tenuous relevance of Jacques Derrida', and concerns his views of ethics and responsibility. In his overview of CSR, De George attributes differences in conceptions of CSR to differences in context, not to differences in our views ethics and responsibility. Indeed for him, '[t]he function of ethical theory... is to make sense of our *common* human and individual moral experience' (78; my italics), and in this regard, corporate responsibilities reflect the demands of society 'to the extent that they are not ethical or moral responsibilities' (76). De George's position is however only tenable if one accepts the rigid distinction between the normative and descriptive categories, since this is the only way in which it makes sense to speak of our common morality, whilst simultaneously allowing for contextual differences. De George's position is indicative of traditional conceptions of CSR, and what is lacking in these conceptions is a *critical* reflection on (as opposed to merely a comparative account of) how our theories and embedded practices shape our views of morality and responsibility (as enacted in CSR practices).

In this regard, it is useful to again recall Raffoul's (2008: 271) critical reflections on the current and growing development of 'applied ethics', which, he argues, 'is paradoxically accompanied by a peculiar blindness regarding the nature of ethics, and a neglect of a genuine philosophical questioning concerning the meaning of the ethical'. A deconstructive complex ethics can therefore help to raise anew the question of ethics, and draw attention to the exclusions, biases, and limitations that define our CSR models. As argued in this study, this is imperative in promoting responsible action in a complex, uncertain, and global world.

Ideas pertaining to our common human and moral experiences are also based on a notion of responsibility that is both causal and intentional. All the standard ethical theories work with these premises, and are therefore agent-centric, since the emphasis lies on the agent performing the action, not on the patient or object on whom the action is being performed (Siponen 2004). This notion of responsibility also informs our understanding of corporate responsibility (where either corporate agents or the corporation itself is viewed as rational and intentional), and where – as will be illustrated below – responsible action remains trapped within a dialectical process. Within this process, responsibility becomes a mechanism by which to maintain an equalising justice. This limits the meaning and impact accorded to the notion of responsibility. Contrary to this traditional conception, a Derridean notion of responsibility pushes against the limits of the dialectic, in an attempt to exercise a true responsibility that transcends the reciprocal demands and expectations of a circular economy that currently inform our understanding of CSR. This insight can also be restated as follows: contrary to traditional conceptions of responsibility that are defined within a restricted economy, a deconstructive notion of responsibility cannot avoid the general economy. The reason for introducing the notions of a restricted and general economy (which are discussed in more detail below at the hand of Derrida's work) is because these concepts are important in further elucidating the nature of Derrida's ethics, and for defining his ethics against traditional conceptions of ethics that inform our understanding of responsibility.

The Restricted vs. the General Economy

In order to introduce the discussion on the restricted and general economy, it is useful to recall Morin's (2007) distinction between restricted and general complexity. Whereas the former notion is premised on the possibility of discovering rules and laws that can help us to decipher the complexity, the latter understanding is encapsulated in frameworks that attempt to work with the irreducible nature of complexity that characterises our understanding of the world and our lived experiences. By extension, thinking on the limits of responsibility within the framework of the general economy implies a serious engagement with the complexities that inform the very meaning of responsible action.

The (Im)possibility of the Sovereign

As noted above, the notion of a general versus restricted economy also features in Derrida's writing, particularly in his work on Bataille's engagement with the Hegelian dialectic (see Derrida 1978), and his work on the Aristotelian distinction between the economic and an-economic (see Derrida 1992a, 1994, 2002a).

With regard to the former, Derrida shows how the Hegelian dialectic – which is brilliantly described in Hegel's discussion on the relation between the master and the slave in the *Phenomenology of Spirit* (1977)[2] – is always interrupted by a third term, which Bataille (1947) calls the sovereign. The role of the sovereign is to disrupt the dialectic between master and slave by engaging with the excess of meaning (i.e. the complexity that exceeds calculation and complete knowledge) that exists within the restricted economy itself, and that cannot be assimilated into the Hegelian synthesis or *Aufhebung*. In other words, the sovereign draws attention to the place of the general economy within the restricted economy.

The sovereign should also be understood in terms of the *aporia* of possibility and impossibility. Derrida (1978: 256) describes the sovereign as something 'totally other' and 'as an excess outside of reason' (255). In other words, the sovereign (like all Derrida's limit concepts) is impossible, because for it to appear, it would need to be drawn back into the dialectical process, and back into the horizon of meaning and knowledge. Again, the impossibility of the sovereign should not be understood as the modal opposite of the possible, but rather as the condition for the possible. This is because the sovereign serves the important function of limiting and opening up the restricted economy, by making 'the seriousness of meaning appear as an abstraction inscribed in play' (256). The sovereign thus manifests as the traces that pervade the restricted economy and that prevent the Hegelian *Aufhebung* from playing out, thereby frustrating the quest for perfect (ethical) knowledge.

Chrematistics: Beyond the 'Ethicality of the Proper'

The idea of the restricted and general economy can also be conveyed in terms other than the master-slave dialectic and sovereignty. In 'Counterfeit Money' (1992a), Derrida calls on the distinction between the economic and the an-economic – which has its origin in Aristotle's (1981) *Politics* – in order to present us with another way of thinking about this difference.

In Aristotle, the sphere of the economic is the sphere of the proper or the *oikos* (what Derrida (1994: 26) calls '*la cloture économique*'). It is the closed sphere of the home and the hearth, wherein a virtuous relation with the economic is possible. Such a virtuous relation is established through practicing an equalising justice (Critchley 1999). The an-economic, on the other hand, represents the improper to ethical life i.e. money-making or chrematistics. This is because it refers to a situation in which 'the use value of an object has been exceeded by exchange value, that is to say, when the de(con)structable infiniteness of accumulation and desire has been introduced into the finite circle of the *oikos*' (168). Once open-ended desire is introduced into the *oikos*, the calculable and symmetrical structure of an equalising justice is ruined. This is because, as Derrida (1992a: 158–159)

[2] For a good introduction see Kojève (1969).

notes, the ethically proper is contingent upon limits, which can be controlled and mastered, whereas the ethically improper marks a transgression of these limits: 'for Aristotle, it is a matter of an ideal and desirable limit, a limit between the limit and the unlimited, between the true and finite good (the economic) and the illusionary and indefinite good (the chrematistic)'.

In 'On the "Priceless"', Derrida (2002a) argues that whereas the economy of the home and the hearth is intimate and known, the market economy is vast, and unknown, resulting in 'consequences of the incalculable' (321). One consequence resulting from the transgression of the *oikos* and the family is market speculation, which Derrida (321) defines as 'the labor without labor of capital, the accumulation, the fetishism of the commodity and the monetary sign'. Whilst this is by no means a positive development, chrematistics, in ruining the ethicality of the proper, also creates the possibility for conceiving of an ethics that transcends the proper. In other words, just as the sovereign can transcend the master-slave dialectic, in order to introduce play, possibility, and space within the system; so too chrematistics (in creating infinite possibilities and incalculable consequences) offers 'the chance for any kind of hospitality... the chance for the gift itself. The chance for the event.' (Derrida 1992a: 158; also see Derrida 2002a).

Critchley (1999) explains this citation as follows: money (in disrupting the restricted economy, or in passing beyond economic calculation) opens up the possibility for a form of an-economic giving, for a donation without return (the opposite of equalising justice), and therefore for the possibility of the gift or of an irreducible justice. For Derrida (1992a), the possibility of the gift is marked by its very impossibility. John Caputo (1997b) explains that within a circular economy, a gift begins to annul itself as soon as it is given, because a reciprocal expectation is created when I say 'thank you' for the gift. Gifts are therefore more-or-less economic transactions, even though they do not appear as such. Pure gift-giving can only take place beyond intentionality, where one neither intends nor expects to give or receive a gift. Caputo (144) writes that, for Derrida, the only way out of this *aporia* – this pollution of the gift – is 'to push against this limit, to transgress this boundary as far as possible, or (im)possible, to make a passage to the limits, to embrace this impossibility, to try to do the *im*-possible, which is not a simple logical contradiction.' In other words, true gift-giving – like justice, responsibility and ethics – must take place beyond the calculable programme, i.e. beyond exchange, circulation, economic circularity, and gratitude (Derrida 1992b), and therefore cannot avoid chrematistics.

Implications

Two important insights can be taken from Derrida's discussion on the sovereign and chrematistics. Firstly, the restricted economy is always already interrupted by a general economy that resides within the heart of the restricted economy itself and that serves to disrupt the logic of this economy. Secondly, responsible and ethical action should be conceived of in terms of the general economy, which – as the expression of the impossible – serves to possibilise the possibility of a more radical notion of responsibility and ethics. In other words, the impossible possibility of the general economy ruins an equalising notion of responsibility and the ethicality of the proper, and thereby offers up these concepts for genuine philosophical questioning.

Following these two insights, we can thus argue that a Derridean-informed notion of corporate responsibility alludes to an understanding of responsibility that transcends the calculable programme of the capital-labour dialectic. If corporations only practice responsibility within the limits of the dialectic, responsibility becomes no more than the practice of corporate duties and responsibilities, as determined by society, in exchange for certain corporate rights and privileges. Responsibility, on this take, would be solely determined by the social contract, which stipulates the terms for an equalising justice or circular exchange between societal interest groups and economic interests groups.

Although this view of responsibility informs dominant notions of CSR, Derrida has a much more radical notion of responsibility in mind – one that definitely introduces a new category to the CSR debate. To summarise: Derrida's notion of responsibility provides a challenge to 'responsibility' conceptualised within the framework of the status quo, by drawing attention to the incalculable possibilities and consequences deriving from different notions of responsibility. Such a re-inscription of responsibility is certainly not premised on the idea of the *oikos* and a common humanity, but rather represents an attempt to think beyond the possible, and view responsibility as a demand without exchange, without recognition, and without rules. On this view, responsibility becomes an expression of ethical complexity – something which, in practice, always pushes the limits of its own expression. Derrida's notion of ethics and responsibility is certainly a lot more radical than the traditional conceptions of these terms, and has therefore also met with a lot of criticism, as explained below.

The Tenuous Relevance of Jacques Derrida? Two Charges Against Deconstruction as Basis for CSR

The main virtue of an equalising notion of ethics and responsibility is that it allows for the clear articulation and evaluation of corporate responsibilities. In other words, it allows for 'explicit CSR', and legitimises CSR practices through the promulgation of rules, norms, laws, best practices, and through the exercise of normative pressure by educational and professional authorities. Propagators of this notion of justice, such as De George, are critical of a complex and deconstructive ethics, precisely because the type of responsibility promoted in this view is not seen as *practically* useful. This is because our very definition of responsibility is constantly under deconstruction, and therefore does not provide a substantive basis from which to judge how corporations fare in terms of meeting their responsibilities. Otherwise put, it is not clear on what basis moral judgement can take place. For the purposes of this study, this is termed *the charge of relativism.* A second charge against a Derridean notion of responsibility is defined in this study as *the charge of irreducibility.* This charge can be described as follows: if responsible action can only take place in the moment of the event, and if ethics is irreducible, undecidable, non-subsumptive, and enacted in the singular relation with the Other (as proposed by Derrida), then it is unclear of what value business ethics in general, and CSR in particular, can be to business practitioners.

Both the charge of relativism and the charge of irreducibility are implied in De George's (2008) critique of a deconstructive ethics. Moreover, in this study De George's view of business ethics is treated as paradigmatic of the traditional approach to business ethics. For these reasons, the charges against Derrida will be elucidated at the hand of De George's critique against deconstruction, and the defence of Derrida's views on responsibility will be offered in critical contrast to De George's own philosophical position.

The central arguments that will be put forward in this analysis are as follows: Firstly – and with reference to the charge of relativism – it will be demonstrated that a deconstructive notion of ethics draws attention to the complexities that define our contexts, and thereby highlights that which is excluded from our conceptual frameworks. This is not a relativist position, but rather denotes a modest stance, geared towards openness towards otherness. Furthermore, this modest stance certainly does not mean that our evaluation of decisions and actions are not subject to rules, standards, or norms. Rather, the point is merely that these evaluative categories cannot be exhausted in terms of their conceptual content. Secondly – with reference to the charge of irreducibility – it will be shown that the irreducible nature of ethics is a consequence of the complexities and over-determinations with which we have to grapple, which means that responsible action must be assumed in a specific context; it can never be prescribed beforehand. This argument does not nullify the necessity of business ethics tools; it just limits the status accorded to these tools. It will also be argued that the ethical relation (which Derrida construes as a relation with the singular Other) is always interrupted by a third. In practice, this means that the ethical relation is always accompanied by a political dimension, which disrupts the ethical moment, and which demands business strategy and institutionalisation.

The Charge of Relativism

De George (2008) argues that a serious problem with Derrida's deconstructive reading of responsibility is that it neither allows us to measure responsibility, nor relates it to praise, blame, shame, punishment, or any other notion with which it forms a conceptual net. Consequently, he concludes that '[i]t is not clear that Derrida recognizes any objectively right action, and hence one is always unsure because there is nothing to be sure about' (82). De George (83) also links this criticism to both Derrida's anti-foundational ethics (arguing that '[b]ecause he questions foundationalism in any aspect of thought, he questions the foundations of morality'), and to the charge of irreducibility, since if our decisions can only be taken beyond knowledge, then it becomes impossible to judge the merits of these decisions. In other words, De George argues that, '[n]either a corporation's actions nor the actions of individuals acting for a corporation fall under any ethical rules'(84). Therefore, the above criticisms of a deconstructive ethics offered by DeGeorge support and expand on the charge of relativism. He further illustrates the practical consequences arising from this interpretation of Derrida's position in the following example:

> It appears that on Derrida's view, if a corporate manager decides to pay much more than the going wage in a less developed country and to ensure safe and healthy working conditions, that decision is not one that the manager can feel secure is better than the alternative of running a sweatshop. And if he does feel secure, he is mistaken. If that is the proper conclusion, then those in CSR would be understandably puzzled (82).

Two Counter-Examples

Global Standards for Responsible Manufacturing

As with De George's example cited above, the first example offered in defense of Derrida's position also pertains to labour practices and CSR, and is taken from a *TIME* article, entitled 'Manufacturing: the burden of good intentions' (Power 2008). This article addresses the issue of whether the same labour standards should be applied the world over. Specifically, it is argued that the "comply-or-die" model to labour rights and corporate social responsibility can actually hurt workers. This is particularly true for workers in non-Western countries, where manufacturers are pushed to develop products at increasing quicker and cheaper rates, whilst still maintaining Western standards of labour practice.

Carla Power argues that 'living up to CSR's high-minded ideals is proving extraordinarily hard in countries like China and India that are at the heart of global manufacturing.' She describes how, in the face of increasing global competition, Chinese suppliers set up "five star factories" that comply with international labour standards, but that these factories run alongside "shadow" factories which operate to meet actual (often Western!) order deadlines. Citing Michael Kobori, head of supply chain social and environmental sustainability at Levi Strauss, Power argues that '[t]he craze for auditing has, paradoxically, led factory owners to create such ["shadow"] factories. It also sops up resources that could be channeled toward improving labor conditions.'

Mukhtarul Amin, managing director of Superhouse Ltd., an Indian clothing manufacturer with clients such as Esprit and Diesel, candidly states that he, too, cannot meet all his CSR obligations. Some, he says, are too difficult to meet, whereas others are irrelevant for Indian society. One retailer expects of their suppliers to have official documentation of workers' ages, but many rural Indians have no such documentation. Acquiring the documentation is a time-consuming and costly exercise, which eats into company profits. In order to offset these extra costs, companies often engage in shrewd labour practices, such as employing additional short-term workers instead of paying overtime.

In yet another example cited by Power, managers who refuse to let migrant workers, who wish to return home as soon as possible, work illegal overtime, risk losing workers to less stringent factories. All these examples lead Rosey Hurst, founder of Impactt, an ethical trade NGO based in London, to succinctly summarise the lesson that we can learn from this comparative study: 'The world is a complicated place'. What these examples show is that context is important for understanding our responsibilities, and that one should be weary of imposing 'universal' notions of CSR on developing economies, since the effects can leave workers worse off, rather than better off.

Global Compact

Niklas Egels-Zandén and Markus Kallifatides (2009) make a similar point in their critical examination of the influence that the Enlightenment tradition exercises on our institutional models for global corporate social responsibility, specifically the Global Compact. The Global Compact was launched by the United Nations in 2000, and seeks to promote private sector compliance with ten basic principles covering human rights, labour standards, the environment, and anti-corruption. Egels-Zandén and Kallifatides argue that the principles promulgated in the Global Compact are either linked to human rights issues that are rooted in European and U.S. ideals that pertain to a liberal notion of society; or can be described as technocentric in nature. Both the human rights tradition and technocentricism are founded in the Enlightenment Ideal, wherein human rights, rationality, reason and science are viewed as the mechanisms for conquering ignorance, religion, oppression and superstition.

This analysis is conducted at the hand of a case study investigating a rural electrification project undertaken in Tanzania by Asea Brown Boveri (one of the largest engineering companies in the world) under the aegis of the Global Compact. What is specifically analysed are the aspects of the local Tanzanian context that are challenged and changed by the values of the Global Compact. In this regard, changes brought about by the project included a weakening of the position of 'traditional medicine', a weakening of the position of influential villagers, a shift towards an emphasis on 'know-how', and a strengthening of the position of women in the village. As a result of the electrification project, the village's institutional environment is now better aligned with the Global Compact principles. However, this change did not happen without conflict. Due to this conflict, Egels-Zandén and Kallifatides argue that changing the rules of the game may not be to everyone's benefit, and may similarly lead one to question the universality of human rights, particularly in non-individualist or tribal settings. Furthermore, they note that certain operational aspects of the project were in direct contravention with the Global Compact principles. For example, using diesel to fuel electricity was at odds with the environmental principles, using children to dig trenches was at odds with anti-child labour principle, and the administration of unofficial payments went against the anti-corruption principle. Yet, Egels-Zandén and Kallifatides were surprised to note that none of these challenges led to changes to the global institutional environment. In fact, these challenges were not even reported at this level due to the conscious and unconscious filtering of information from the ground up, and because none of the local role-players communicated with the international stakeholders. In other words, the international manager acted as a gatekeeper, by controlling information and stakeholder access.

Egels-Zandén and Kallifatides attribute the reasons for this to the fact that many influential stakeholders (including NGOs, the media, and consumers) actively work to ensure business compliance with the Global Compact. These stakeholders are influential because they can do reputational damage to firms, and there is therefore much pressure on firms to appear compliant. Furthermore, these firms are too small to challenge the institutional environment, and in the global setting they are defined as value-takers as opposed to value-setters. On a local level however, the tables are turned, since the villagers are not in a position to challenge big companies (especially when these companies bring them the gift of electricity). Therefore, on a local level the firm acts as the value-setter, and the villagers are the value-takers. Egels-Zandén and Kallifatides conclude that the result is that some local traditions are dismantled, whereas the so-called international principles remain unchallenged. There is thus a tension between 'universal' standards and local practices, and the desirability of the shift to these 'universal' standards is by no means evident.

This *Given* Ethics: Acknowledging the Ethics of Complexity

This latter case constitutes an apt example of the naturalisation and universalisation of certain value positions, and it is here where a Derridean notion of responsibility and a deconstructive ethics prove invaluable, as both offer a strong challenge to, and forces us to 'reconsider in its totality, the metaphysico-anthropocentric axiomatic that dominates the thought of just and unjust' in the West (Derrida 1992b: 19). In this regard, Derrida is very cautious of the conceptual nets of which De George speaks, because in deciding how to measure responsibility, or how to assign praise or blame, we 'reproduce, under the guise of describing [a certain ethicality] in its ideal purity, the given ethical conditions of a given ethics' (Derrida 1988: 122).

This given ethics however, can never be equated with a universal ethics, and doing so excludes certain conditions no less essential to '*this given* ethics or of *another*, or of a law that would not answer to Western concepts of ethics, rights or politics' (122). The task of deconstruction is therefore to challenge the 'very concept of responsibility that regulates the justice and appropriateness (*justesse*) of our behaviour, of our theoretical, practical, ethico-political decisions' (Derrida 1992b: 20). Derrida (1988) further argues that in employing a deconstructive ethics, we not only draw attention to that which has been excluded, but we also create an opening for thinking anew about responsibility and the concepts related to responsibility – which include 'property, intentionality, will, freedom, conscience, consciousness, self-consciousness, subject, self, person, community, decision, and so forth' (Derrida 1992b: 20). In trying to account for other conditions of responsibility, one might even 'open or recall the opening of another ethics, another right, another "declaration of rights," transformation of constitutions, etc.' (Derrida 1988: 122). From the above, it is clear that – far from resulting in a relativist position that allows us to say nothing about responsibility – Derrida's deconstruction calls for an increase in responsibility!

Again, it is useful to bear in mind the insights gleaned with regard to modelling complex systems. The descriptions that we accord to our models (including our models of CSR) are contingent on the resources at our disposal, and often reflect our prejudices and biases. As such, these models are unlikely to account for different conceptions of rights and responsibilities. Therefore, the only way in which we can do justice to different ways of being, is to subject our models to critical scrutiny in an attempt to recall the opening of another ethics, of which Derrida speaks. In other words, ethics demands that we transgress our current conceptions of CSR, and try to imaginatively seek out other alternatives in an attempt to enlarge our view of responsibility.

De George's attempt to define contextual differences in terms of non-moral responsibilities represents an effort to define a *given* ethics as *the* ethics, or the only way of being. Thus, De George does not take sufficient cognisance of the provisional or limited nature of our ethical models. This is not uncommon. In 'Ethics, Institutions, and the Right to Philosophy' Derrida (2002b: 9) states that we should go 'beyond the old, tiresome, worn-out, and wearisome opposition between Eurocentricism and anti-Eurocentricism.' Derrida (9) argues that one way in which to achieve this is to '*take into account and de-limit* the assignation of philosophy to its Greco-European origin or memory.' This neither means that we should reaffirm this history, nor discard it (indeed we can't!), but that there must be:

> the active becoming-aware of the fact that philosophy is no longer determined by a program, an originary language or tongue [in De George's terms, 'a common humanity'] whose memory it would suffice to recover so as to discover its destination (10).

In Defense of Derrida: Overcoming the Charge of Relativism

Being attentive towards different ways of being and philosophising is not a relativist stance, but a modest stance, geared towards openness for otherness. As stated before, Derrida (1999a) firmly denies that he is a relativist (if by relativism, we understand a doctrine with its own history of denying absolutes). Derrida concedes that he refuses to reduce differences, but does not understand this to be an example of relativism, but rather an example of discernment. In this regard, he asks:

> Are you a relativist simply because you say, for instance, that the other is the other, and that every other is other than the other?... If I want to pay attention to the singularity of the other, the singularity of the situation, the singularity of language, is that relativism (78).

This argument also does not imply that one should forego laws, norms, and rules, as presumed by De George. In response to this line of critique, Derrida argues that we *need* laws, rules, and norms. In fact – as will shortly be demonstrated – the deconstructable nature of law is exactly that which gives meaning to law or to our concepts (moral or otherwise). As argued in Chap. 2, a structural condition of concepts is that they must be formulated according to the logic 'all or nothing' (Derrida 1988). In other words, every time I utter a word or a phrase, I do so at the exclusion of all other words and phrases. Even vague statistical approaches cannot overcome this problem: relativism (as a concept) is just as determinate in its logic as absolutism (as concept) is. This is the nature of conceptual language: it is determined by rules and laws, and Derrida uses the example of a traffic light to show that the law is often enabling: 'the essence of law is not necessarily tied to negativity (prohibition, repression, etc.)' (133).

However, as should be clear from the discussion, the task of deconstruction is to *challenge* law in the name of justice and ethics. This is also not to say that our challenge to concepts of law, social responsibility, or business ethics cannot be a strong challenge. Derrida is quite clear in stating that deconstruction does not claim that there is no better or worse. In this context, consider the following citation, also taken from the 'Afterword' (146), where Derrida again responds to his critics' charge of relativism:

> For of course there's a "right track" [*une "bonne voie"*], a better way, and let it be said in passing how surprised I have often been, how amused or discouraged, depending on my humour, by the use or abuse of the following argument: Since the deconstructionist (which is to say, isn't it, the sceptical-relativist-nihilist!) is supposed not to believe in truth, stability, or the unity of meaning, in intention or "meaning-to-say," how can he demand of us that we read *him* with pertinence, precision, rigor? How can he demand of us that his text be interpreted correctly? How can he accuse anyone else of having misunderstood, simplified, deformed it, etc.? In other words, how can he discuss, and discuss the reading of what he writes? The answer is simple enough: this definition of the deconstructionist is *false* (that's right: false, not true) and feeble; it supposes a bad (that's right: bad, not good) and feeble reading of numerous texts, first of all mine, which therefore must finally be read or reread. Then perhaps it will be understood that the power and truth (and all those values associated with it) is never contested or destroyed in my writings, but only reinscribed in more powerful, larger, more stratified contexts. And within interpretative contexts... that are relatively stable, sometimes apparently unshakable, it should be possible to invoke rules of competence, criteria of discussion and of consensus, good faith, lucidity, rigor, criticism and pedagogy.

Far from endorsing relativism, the deconstructionist is intimately involved with truth, which Derrida (1999a: 77) defines as 'the quality of a statement, a judgement or an intuition related to something which you might call a fact, but truth is not reality'. The search for a better truth is not a search for absolutes or a pure science, but a way of taking into account difference, or of 'negotiating with alterity' (Wood 1999: 109). In this regard, David Wood (110) writes that 'deconstruction "itself" and, indeed, the concept of "responsibility," to which Jacques Derrida has recently given so much weight, are each nothing *other* than experience regained. Deconstruction is, if you like, the experience of experience.'

Derrida's project – which focuses on different (better) ways of being – is at odds with the traditional way of doing business ethics (as exemplified by De George's position), which is essentially a way of downplaying differences in the name of a common ethical experience, or a common moral foundation. Derrida (1999a: 74) readily admits that his is a 'poor work' when it comes to spelling out 'this is what you have to know' or 'this is what you have to do'. But the important point here is that this does not imply that we can't know or do anything. Wood (1999: 105) affirms this when he writes that '[w]e owe to concepts like "justice," "rights," "duty," "virtue," "good," "responsibility," and "obligation," our capacity for ethical judgement' but warns that 'the work of clarifying and codifying the scope and significance of these terms is the source of another danger – the calculation of our responsibility, in which the ethical as an openness to the incalculable is extinguished.'

To guard against this, we should deconstruct and employ the logic of 'concepts' such as *différance*, iterability, supplement etc. We do this not in the name of relativism or indeterminacy, but in the name of ethics, and when we do this, Derrida (1988: 117) argues that 'it is *better* to make explicit in the most conceptual, rigorous, formalizing, and pedagogical manner possible the reasons one has for doing so, for thus changing the rules and the contexts of discourse.' Therefore, as demonstrated in this analysis, those who view Derrida's position as relativist show a poor understanding of his work. However, in order to demonstrate the value of Derrida's ethical position, it is also necessary to address the second charge against him, namely the charge of irreducibility.

The Charge of Irreducibility

The charge of irreducibility stems from Derrida's view of the ethical as an experience of the (im)possible, which is further highlighted in his work on undecidability and the ethical relation with the Other. According to De George (2008), Derrida's emphasis on, and understanding of, these notions translates into a undecidable and altruistic notion of justice that cannot inform our understanding of corporate responsibility, which is necessarily linked to the economic and political spheres. Although undecidability is a necessary condition for Derrida's non-subsumptive view of responsibility and the ethical relation, these two dimensions (which both reinforce the charge of irreducibility) will be treated separately in the analysis below, in order to demonstrate why this second charge against Derrida also does not hold.

Undecidability and the Charge of Irreducibility

As mentioned above, De George (2008) takes specific issue with Derrida's notion of undecidability, arguing that if (as claimed by Derrida) undecidability is a condition for ethics, politics, justice, and responsibility, then – far from bringing us closer to fulfilling our responsibilities – we are left hopeless to act. For De George, '[t]he unsettling aspect of the act of deconstruction... is that we seem never to get to an answer, and that whenever we arrive at an answer we are assured that it must be wrong' (82). Given this interpretation, he concludes that a deconstructive mindset 'reduce[s] those in business, who have to make decisions, or their critics, to the position of an undecided Hamlet' (82). In order to overcome this critique, it must firstly be demonstrated why we cannot do away with undeciability when engaging with our ethical responsibilities, and in this regard, Derrida's (1992b) discussion on the relation between justice and law, as provided in his illuminating text entitled, 'Force of Law: "The Mystical Foundation of Authority"', proves helpful in furthering the analysis.

The Interplay Between Justice and Law

In 'Force of Law', Derrida (1992b: 17) argues that justice is incalculable and 'must always concern singularity, individuals, irreplaceable groups and lives, the other or myself *as* other, in a unique situation.' In contrast to the incalculability of justice, 'law is the element of calculation'; it is the 'rule, norm, value or the imperative of justice which necessarily have a general form' (16, 17). Whereas law is deconstructable, justice is not, which also means that justice is impossible; or, 'an experience of the impossible' (16).

As stated above, De George (2008) understands this to mean that Derrida does not recognise objectively right actions, which means that we cannot judge whether a moral risk has paid off or not. Yet, De George argues (with reference to arguments made in Jones et al. (2005) *For Business Ethics*) that, despite claiming not to know what justice is, those who support a Derridean view of justice (such as Jones and his co-authors) can confidently discern between just and unjust actions and 'somehow they, like the rest of us know that stealing from pension funds, false accounting, sweatshop labour, manipulative marketing, and a host of other practices are wrong' (De George 2008: 83). How – he muses – is it possible to condemn these practices whilst claiming that justice is an experience of the impossible?

Derrida (2002c) argues that justice, as a limit concept, is a way of saving the legacy of philosophy, which we cannot do without (including all our metaphysical notions such as the *a priori,* origin, or foundation). However, the work of limit concepts lies in 'rethinking the meaning of the "possible," as well as the "impossible," and to do so in terms of the so-called condition of possibility, often shown to be the "condition of impossibility"' (354). In other words, the status of justice as a limit concept safeguards justice (the condition of impossibility), whilst simultaneously allowing for its inscription in the moment of the event (the condition of possibility). Justice can never be encapsulated in a rule or norm. However, justice also needs rules. As Derrida (1999a: 72) puts it: 'Justice requires the law. You can't simply call for justice without trying to embody justice in the law.' Therefore, justice demands that we interpret and judge the rules that we apply, each time we apply them. In this regard, Derrida (1992b: 23) writes: 'No exercise of justice as law can be just unless there is a "fresh judgement"', which means that we must take responsibility for the rules that we employ. Justice cannot amount to the application of a rule (a position known in law as 'legal positivism'), since in such circumstances, there is no responsible action.

It is, as previously argued, useful to think of Derrida's conception of justice as constituting an expression of ethical complexity. Since we cannot understand complexity in its fullness, we are forced to reduce it through modelling. Our models of justice therefore constitute law. Although we try to remain faithful to this ethical complexity, law is always a partial and incomplete model of justice, and should therefore continually be revised, hence the fact that justice and law are tied together in 'the self-contradictory relation of unrevisable separation and permanent provocation' (Teubner 2001: 42).

Hamlet as a Paradigm of Action

If we translate Derrida's discussion of justice and law into the language of business ethics, we can say that ethics and responsible action can never simply be encapsulated in codes and CSR statements. Although business ethics tools are useful for helping us to think through our responsibilities, they cannot be blindly applied. Doing so will result in ethical positivism (which is akin to legal positivism), and which results in a type of procedural ethics that 'robots can learn and copy' (Adorno 1973: 30). Ethical action (like justice) is only possible when we take responsibility for our decisions. This also means that responsibility should be assumed anew every time a decision is made or an action taken. Responsibility (like justice) is therefore singular, and is something that manifests in the event.

De George (2008: 84) is correct in stating that for Derrida, 'one cannot simply be following a rule'. De George however argues that this account of responsible action is at odds with the phenomenology of our moral experience, and for this reason he supports an Aristotelian analysis, where one acts virtuously because one has developed the habit of acting virtuously. The difference between De George and Derrida is therefore that, for the former, responsible action is embodied in the rule or habit (which we, as moral agents, then freely choose to follow); whereas for the latter, responsible action resides in the attempt to act beyond the rule when making an ethical decision. In the words of Raffoul (2008: 284), '[e]thical responsibility is thus a matter of *invention*, not application of a rule.'

Although De George does not spell out the reasons why merely following the rule accords better with our moral experiences, the answer presumably lies in some argument pertaining to our common human morality and our bounded rationality. But, it is precisely because we are boundedly rational, and because the world is a determinate place, that we have to go through what Derrida (1999a: 66) calls 'the terrible process of undecidability', which also alludes to the impossibility of the decision. Derrida's treatment of the question of 'undecidability' (which so many of his critics confuse with 'indeterminacy') amounts to an acknowledgement of our limitations and of our inability to reconcile conflicting demands, rather than denoting paralysis when faced with a decision. Derrida writes that those critics who charge him with saying that the text means anything, usually betray an anxiety with regard to the fact that 'texts *may* call for interpretation... there may be some complication in a text' (79). With regard to the text (and therefore also with regard to context and differences in and between contexts) Derrida (79) writes:

> I would say that the text is complicated, there are many meanings struggling with one another, there are tensions, there are overdeterminations, there are equivocations; but this doesn't mean that there is indeterminacy. On the contrary, there is too much determinacy. That is the problem.

We experience undecidability when confronted with a problem where 'I know that the two determined solutions are as justifiable as one another' and I must make a decision which is 'heterogeneous to the space of knowledge' (66). Real choice (and hence, autonomy) is therefore always constrained by what Morin (2008: 113) calls the 'hazards of *existence*'. Autonomy should be conceived 'not in opposition to, but in complementarity with, the idea of dependence' (114). Complexity is the outcome of freedom and structure, and it is these elements that, together, constitute the process of undecidability, which guards against the decision being reduced to the mere application of a rule. As noted by Raffoul (2008: 284), the 'aporia [of the undecidable] is the condition of decision and the very locus of freedom', and this view is also affirmed in Derrida's own words, when he writes that: 'Where I still have a space for choice, I am in the antinomy, the contradiction, and each time I want to keep the greatest possible freedom to negotiate between the two' (Derrida 1999b: 48, trans. F. Raffoul).

Derrida (1999a: 66) states that '[e]thics and politics, therefore, starts with undecidability', which haunts every decision. The ethical moment is the precise moment where, after I have 'prepared as far as possible by knowledge, by information, by infinite analysis' (66), I must go beyond knowledge, and act. For this reason, the distinction between good and evil (i.e. ethics) cannot depend on knowledge: 'we should not know, in terms of knowledge, what is the distinction between good and evil' (66). It is with reference to the very figure of Hamlet that Derrida turns to explain the importance of going through a process of undecidability. In this regard he writes, 'if we assume that Hamlet is a figure of paralysis or neurosis because of undecidability, he might also be a paradigm for action: he understands what actions should be and he undergoes the process of undecidability at the beginning' (68).

What is therefore again reiterated in this analysis is that Derrida supports a non-subsumptive view of responsibility. Although calculation, law, and substantive ethical theories are necessary, these tools can never *give* us ethics. Decision and judgement is needed for responsible action to occur. Responsibility means *assuming* responsibility and taking a decision in the face of the undecidable.

The Ethical Relation and the Charge of Irreducibility

Levinasian Responsibilities

The discussion on a Derridean notion of responsibility would be incomplete without reference to Levinas, who greatly influenced Derrida's own views on the matter. Levinas understands ethics as prior to moral questioning, and *a fortiori* to moral law. In other words, Levinas is first and foremost concerned with 'the primordial ethical experience', from which certain moral questions, maxims, and judgements may be derived (Critchley 1999: 3). Christina Howells (1998) argues that, for Levinas, the ethical experience begins with recognising the otherness of the Other, which takes place in the face-to-face encounter with the other that serves to challenge my 'subjecthood' or 'myself as ego or as subject with inalienable priority' (124). It is through challenging or questioning my primacy as subject, that the face of the other frees me from reification, and consequently grounds the ethical. Levinas thus defines ethics as 'the putting into question of my spontaneity by the presence of the other' (Levinas 1969: 43).

Derrida takes up, refines, and extends this Levinasian view of responsibility, as an event that calls the subject into question by demanding a response to the needs of the Other (Jones 2007). The nature of the event of the Other is also explained by Derrida with reference to the figure of Hamlet, which Derrida (1994) recalls in the opening pages of *Specters of Marx*. Of particular interest for Derrida is Hamlet's feeling that 'the time is out of joint'. Critchley (1999) argues that, in his analysis of Hamlet, Derrida weaves together Hamlet's experience of disjuncture, with the asymmetry of the ethical relation in Levinas, as well as with Martin Heidegger's meditation on time and justice. According to Critchley, what interests Derrida is Heidegger's attempt to think justice in relation to time. Contrary to how Heidegger would have it, Derrida (1994) claims that there can be no presencing of the present, in terms of a jointure or a donation of time, in which we experience the phenomenality of the Being of beings as disclosure or unconcealment of what is present. The moment of justice as an event cannot come into presence (be present) in a manner that disrupts the flow of time. The reason for this is that the Heideggerian conception of justice demands absolute symmetry or the revelation of the proper of the Other as presence. In other words, Heidegger supports a logocentric view of justice. For Derrida, the moment of justice resides in the disjoint, in the disjuncture of the ethical relation with the Other. Justice as disjuncture, rather than 'justice under the sign of presence', is the very 'condition for the presencing of the present' (27). Justice (and responsible action as an instance of justice) cannot be assembled into a totality: it can only be experienced in the singularity and asymmetry of the event i.e. in the (im)possible face-to-face encounter with the Other.

De George's Contention

De George (2008) argues that Derrida (in appropriating Levinas's understanding of the ethical relation) supports not only a non-subsumptive, but also a reforming notion of ethics, which is completely altruistic or other-orientated. De George (84) contends that '[c]ommutative justice, which involves exchanging equals for equals, has no place here'. De George's contention is thus that there is no place in society or the economy for a Derridean notion of responsibility. Whilst he finds Derrida's radical view on ethics laudable, he argues that it certainly does not constitute our ordinary understanding of ethics. This, he argues, forces us to ask:

> what is the relation that we previously thought constituted the ethical? Do we have to give it up? Do we rename it? Or contrary-wise, why not retain the term 'ethics' for the traditional meaning and call this new reformed relation by some other name, for instance, what is ordinarily thought of as an unconditional relation of love for another? (84)

Derrida never wrote on love, but one can nevertheless ask (along with De George) what consequences Derrida's view of a unique and singular relation with the other may hold for the practice of business ethics generally, and CSR specifically. In other words, the question is: how is ethical action in business possible given the fact that justice and responsibility (as a relation with the Other) remain irreducible, undecidable, and non-subsumptive? De George's contention amounts to the fact that a Levinasian view of responsibility is applicable only between individuals, yet '[t]he task of CSR is a different task, namely influencing those in business to act in a way that is more positive in its effects on human beings, on the environment, on the common good than is often the case' (82). If the ethical relation is constituted by the meeting with the singular other, then the term 'business ethics' becomes an oxymoron. This is because, as De George argues, Derrida's ethical relation is incompatible with the logic of organisations, defined as profit-making entities. The task at hand is therefore to demonstrate that Derrida's view of the ethical relation does not leave us in an impotent position, where corporate responsibility is individualised to the extent that one cannot account for the interests and claims of different and competing stakeholder groups.

The Face of the Third

In order to attempt to address this question, we take as a starting point Critchley's (1999) discussion on deconstruction and pragmatism, in which he questions whether Derrida is a private ironist or a public liberal, which are the categories employed by Rorty (1989) to designate the private and public spheres. Critchley (1999) argues that despite attempts by Platonists, Christians, Marxists, and Kantians to reconcile the private and public domains, Rorty maintains that they remain theoretically irreconcilable. As an example, self-actualisation, self-interest, and personal autonomy are fundamentally incompatible with justice, charity, and love for one's neighbour. Whereas the ironists 'faces up to the contingency of their most central beliefs and desires', liberals are people who are appalled by human suffering and cruelty. The heroine, for Rorty, is the figure of the liberal ironist: 'someone who an committed to social justice and appalled by cruelty, but who recognizes that there is no metaphysical foundation to her concern for justice' (85).

Critchley's (98) hypothesis is that if we accept Rorty's definition of the public domain and if 'deconstruction is ethical in the peculiarly Levinasian sense... then deconstruction would be concerned with the suffering of other human beings and would therefore qualify as public by Rorty's criteria.' It is precisely *because* deconstruction is concerned with the irreducibility of the ethical relation that it has significant practical consequences.[3] Critchley however argues that Derrida understands justice in an ethical (as opposed to a political) sense, but that this understanding doesn't make justice apolitical. To explain this, it is necessary to follow Critchley in his analysis of the Levinasian understanding of justice.

In *Totalité et Infini,* justice functions as a synonym for ethics, understood as '*la relation avec autrui*' (the relation with the other) (Levinas 1990: 62). It is this specific understanding of justice that Derrida (1992b) refers to in 'Force of Law'. However, in his later works, Levinas differentiates between this understanding of justice and justice as distinguished from the ethical relation. Critchley (1999: 100) paraphrases Levinas's understanding of justice in *Autrement qu'être ou au-delà de l'essence* (1974) as a question that 'arises when a third party arrives on the scene, obliging one to choose between competing ethical claims and reminding one that the ethical relation is always already situated in a specific socio-political context.' The reason why Critchley claims that Derrida accepts an ethical (as opposed to a political) notion of justice is because, for Derrida, the experience of justice (or ethics) cannot be encapsulated in law. Justice is the moment of the decision, or, following Søren Kierkegaard, the madness of the decision (Derrida 1992b).

[3] Although Critchley (1999) makes a convincing argument, which bears interesting implications for this analysis, it is doubtful whether Derrida would accept the distinctions that Rorty employs. In 'Negotiations', Derrida (2002d: 31) explicitly states that 'I do not believe in the radical distinction between public and private.' In speaking of negotiation (which is intimately tied to deconstruction), Derrida (17) writes:

> Negotiation is constantly in a state of micro-transformation. *Every day:* this means it does not stop. This also means that between politics – that is, public life – and private life (interests, desires, etc.) the communication is never broken. I do not believe in the conceptual value of a rigorous distinction between the private and the public... In what I write one should be able to perceive that the boundary between the autobiographical and the political is subject to a certain strain.

However, once the decision is made and the action undertaken, we enter into the realm of politics (which is also the realm of the institution, law, or policy).

In 'Hospitality, Justice and Responsibility', Derrida (1999a: 68–69) elaborates on his understanding of the ethical relation. He argues that the face of the third does not appear *after* the confrontation with the face of the Other. The face of the third 'is already involved in the face to face relation as a call for justice.' The face of the third destroys the asymmetrical, but dual, relation that I experience in my confrontation with the Other. This forces me to abandon my unconditional concerns for the Other and engage in comparison and rationality (in other words, I must return to calculation). However, no amount of calculation can resolve this *aporia*. In the moment of the decision, I have to pass beyond knowledge. The ethical relation therefore means that 'I have a relation to the other in his/her singularity or uniqueness, and *at the same time* the third one is already in place. The second one is a third one. 'You are a third one', that is the condition of justice' (69; my italics). On this reading, it is not a matter of choosing between an ethical and a political notion of justice. For justice to take place, I must be both ethical and political, even though I destroy the very conditions of justice in the moment of justice.

Ethics and Politics

This implies that De George's concern that Derrida collapses ethics into an unconditional, but impotent, relation with the other is invalid, because the third always intervenes, thereby politicising the ethical. The implication that this holds for our understanding of CSR is that we should grapple with the contesting and irreconcilable demands of stakeholder groups – each of which make unique and singular claims upon the organisation. However, we cannot grapple with these claims indefinitely. At some point, management must decide (and decide as responsibly as possible) which claims will be honoured and how – even though this decision might constitute a denial of other claimants' demands. This is the impossible nature of ethics and politics. Even if we acknowledge that value judgements and choices are involved in our decisions, this acknowledgement is not enough to guarantee that good will come of what we do, for the simple reasons that we do not carry full knowledge of the complexity of the situation, and the situation itself is characterised by many tensions and overdeterminations.

In 'Ethics and Politics Today', Derrida (2002e) identifies the following determinants of ethics and politics: firstly, ethical and political responsibility command action. Secondly, both demand a thoughtful answer to the question 'What should I do?' – a questioning without limit (which implies vigilance to the question). Thirdly, both ethics and politics demand that we make the decision with the utmost urgency and haste. The time for reflection is always interrupted by a situation. In short, the relation between a here and now are the common requisites of both ethics and politics. The *aporia* or double-bind of ethics and politics is therefore not urgency against non-urgency, but urgency against urgency. As previously argued, this double-bind also provides the impetus for the critical enterprise. It is precisely because we have to act, despite the complexity, that the critical task is so important.

In 'Negotiations', Derrida (2002d: 29) states that the infinite task of deconstruction as negotiation is 'a work of mediation... a to-and-fro between impatience and patience.' Here, impatience and patience are synonyms for urgency and vigilance, and Derrida warns that we need both – not in small measures, but we need them absolutely. Hence, vigilance/urgency and patience/impatience represents another *aporia*, which we cannot do without. Spivak (1994: 26) sums up this point as follows: responsible thought describes responsibility – 'caught in a question necessarily begged in action' – as attending the call of the irreducible contamination of responsibility. Deconstruction positions itself in the fold of this *aporia*, or between, what Derrida (2002d: 25) calls earlier on in this text, 'affirmation' and 'position'. It is also here that we find one of Derrida's strongest calls for political action:

> One must not be content with affirmation. One needs position. That is, one must create institutions. Therefore, one needs position. One needs a stance. Thus, negotiation, at this particular moment, does not simply take place between affirmation and negation, position and negation: it takes place between affirmation and position, because the position threatens the affirmation. That is to say that in itself institutionalization in its very success threatens the movement of unconditional affirmation. And yet this needs to happen, for if the affirmation were content to... wash its hand of the institution in order to remain at a distance, in order to say, "I affirm, and then the rest is of no interest to me, the institutions does not interest me... let the other take care of that," then this affirmation would deny itself, it would not be an affirmation. *Any affirmation, any promise in its very structure requires its fulfilment. Affirmation requires a position. It requires that one move to action and that one do something, even if it is imperfect* (25–26; my italics).

The Possibility of Corporate Social Responsibility

De George's (2008) concern is that, in attending to the call of the Other, a Derridean notion of responsibility cannot further the task of CSR, which (to recall) is to act in a way that is more positive in its effects on human beings, on the environment, and on the common good than is often the case. In response to De George's question regarding the applicability of Derrida's altruistic conception of responsibility, one can argue that the confrontation with the Other (as a moment of decision and of justice) is what constitutes the ethical relation, and that which is traditionally thought to constitute the ethical (i.e. a commutative or equalising conception of 'justice') are the rules, norms, and calculations that precede the ethical moment. What the above discussion also implies in that, in order to serve the good (the *common* good), we cannot escape this singularity, even though the singularity is always transgressed by a situation. De George is correct in thinking that Derrida expects the impossible, because, in order for an event to take place or to be possible – in order for there to be something like responsibility – it has to be 'the coming of the impossible' (Derrida 2002c: 361). The im-possiblity of possibility is not only negative; rather, as Derrida (361) qualifies, 'it *introduces* the possible; it is its *gatekeeper today*.'

Wood's (1999: 117) understanding of Derrida's notion of responsibility sheds light on the meaning of this last statement: responsibility 'is not quantifiably (or even inquantifiably) *large* and therefore not a basis of guilt through failure to live up to it. It is rather a recursive modality, an always renewable openness.' Here again, we can refer back to the critical enterprise. Since we do not have a bird's eye view on reality, we need to continually engage with complexity, take cognisance of our models, and acknowledge when these models are in need of revision and transformation. To reiterate the words of Preiser and Cilliers (2010: 274): 'The ethics of complexity is thus radically or perpetually ethical.' De George (2008) objects that Derrida's view of responsibility means that companies must 'open themselves up to hospitality and risk being taken advantage of by anyone who chooses to do so.' To this we can respond with Wood's (1999: 117) words: 'Openness does not require that one leaves the door open, but that one is always willing to open the door. Responsibility then is the experience of that openness.' A Derridean notion of responsibility therefore constitutes a powerful expression of both the ethics of complexity and the complexity of ethics.

The Timely Relevance of Jacques Derrida

In the 'Afterword', Derrida (1988: 113) asks whether it is certain that we can eliminate the violence and ambiguity of texts, or whether it is 'even certain that we should try *at all costs*?' This single question looms behind Derrida's entire philosophical project, as becomes clear in Elizabeth Grosz's (2000) exploration of violence in Derrida's work. She raises the question of violence in Derrida's work not only with regard to obvious and manifest violence – i.e. street violence, war, discrimination etc. – but also in terms of more subtle forms of violence (which are rarely termed violence), namely the violence that marks the 'domain of knowledge, reflection, thinking and writing' (190). In these domains, we have to account for the primordial or arche-violence, which is the result of the fact that writing and speech – even arche-writing (the original grunt for food, so-to-speak) – already represent a mode of cutting, the 'worlding of the world' (191). In other words, arche-violence marks the word as 'always already split, repeated, incapable of appearing to itself except in its own disappearance' (Derrida, 1976: 112). However, another subtle form of violence concerns reparatory or compensatory violence, which Grosz (2000: 193) defines as:

> the violence whose function it is to erase the traces of this primordial violence, a kind of counter-violence whose violence consists in the denial of violence... This is a violence that describes and designates itself as the moral counter of violence. This is the violence that we sometimes name the law, right, or reason.

Grosz not only argues for the primacy of violence in Derrida's work, but also in politics and ethics. Indeed, it is *because* of the inescapable violence in language and in being, that ethics and politics are such central themes in Derrida's work. Grosz states that – far from immobilising the practices of politics and ethics, or of refusing to provide answers to political or ethical problems (as is often alleged by his critics) – Derrida offers a profound reconfiguration of the ethical and political activity that centers on the question of violence. In this regard, Grosz (190) states:

> What makes Derrida's work at once intensely political and ethical, while he remains acutely aware of the problems involved in any straightforward avowal of one's commitments to political and ethical values, is his readiness to accept that no protocol, no rhetorical or intellectual ploy is simply innocent, motivated by reason, knowledge or truth alone, but carries with it an inherent undecidability and repeatability that recontextualizes it and frees it from any origin or end.

In his writing, Derrida attempts to avoid moral falsehoods, understood as a complacent denial of the violence that we employ every time that we use language, every time that we argue for a specific position, or every time that we proclaim the Truth of our assertions. Language – like our systems of meaning – is limited and exclusionary: in voicing one position, we deny another, which may be no less relevant; this is unavoidable, but what is avoidable is the refusal to acknowledge this. We have a moral imperative to account for the status of our models, since to ignore this is to risk perpetuating an even greater violence, namely the denial of violence in the name of law, right, or reason. This is not to say that we should not develop a body of knowledge or a community of practice. Rather, the point is that we should guard against the naturalisation of meaning that scholarship can bring about. This point is particularly crucial in the context of ethics, because once we naturalise ethics, any conception of ethics that may deviate from the dominant interpretation can easily be construed as patently wrong, or as something which threatens to poison the purity of our common experiences.

Therefore, the most important function of a complex deconstructive ethics is that it initiates a continual and critical reflection on our views of ethics and how they shape our practices (including our corporate practices). Such a reflection is missing from traditional accounts of CSR, which take for granted a notion of an equalising justice, and it is postulated that this constitutes a serious omission. CSR is a complex phenomenon, as demonstrated by Jones's (2007: 522) assertion that 'the current and urgent global struggles' is testimony of the fact that we do not know 'what it might mean to live and work in anything that might meaningfully be called a free economy or a free society.'

Despite the amount of scholarship dedicated to developing our notions of CSR, there is still little clarity on what CSR *should* entail. This is also confirmed by De George's (2008: 83) statement that '[t]o the extent that it has had any success in improving the lot of human beings, CSR is a positive force in the business arena, even if poorly understood by its practitioners, even if rife with irresolvable conflicts, and even if it is always under the process of deconstructing itself.' Instead of applauding our successes and continuing on the same trajectory, it is time (now more so than ever before) to move beyond our current conceptions of CSR in a bid to re-inscribe our understanding of CSR in larger, more powerful, and more stratified contexts, in an attempt to welcome a better future.

A deconstructive, complex conception of ethics goes a long way towards providing us with the tools for achieving this goal. This approach to ethical decision-making is not against employing ethical tools, including codes, normative theories, and governance reports – all of which advance responsible corporate actions. Before making any decisions we also need to consider our moral tradition, including our conceptions of justice, right, virtue, the good, and responsibility. However, we should also recognise that an over-reliance on business ethics tools can numb the moral impulse to the extent that one forgets that ethics demands decision-making, and that each decision should be taken anew to account for the specificity of the situation. We need to challenge and transgress our schemas and models, with the aim of finding a 'better' fit. We should also however recognise that challenging the status quo implies a measure of risk, because in going against tradition, we are negotiating with alterity and the unknown. This is why we should provide rigorous and clear arguments for changing the rules of the discourse when we assume a position, and in so doing, create institutions. However, the moment we create institutions is also the moment that we close ourselves off to the very alterity that we seek to safeguard. This is why we must always deconstruct anew. What this means is that CSR is as much a matter of signing off on policies and codes, as it is of challenging these policies and codes. In other words, the complexity of the problem has to be constantly thought anew.

This point is particularly poignant in the context of the larger sustainability issues that we face today, which are also related to good corporate social responsibility practices. Given the problems currently topping the global agenda, CSR cannot be limited to drawing up policies against fraud, deception, or discrimination (all of which can be quite well circumscribed in a utilitarian, virtue, or rights-based framework). Assuming responsibility also means dealing with the uncertainties and complexities that confront the world today. We are far from clear on how to proceed in the face of problems such as global warming, food security, financial security, or even infectious disease outbreaks. In this regard, the top management of especially big and/or influential companies might benefit from deconstructive and complexity ideas, even though these ideas cannot provide them with substantive answers regarding the best way forward.

In order to be ethical one must assume responsibility for one's decisions, even when this means confronting certain difficult choices, where no single option presents one with the 'correct' choice. Hence the characterisation of ethics as an experience of the impossible: ethics can only ever find its expression in the irreducible singularity that characterises the moment of the decision. If we cannot assimilate otherness into our existing frames of meaning, then the ethical moment will involve a measure of disjuncture and undecidability. But such a moment is also the moment of vigilance. In today's global world, our actions have far-reaching effects, many of which can only be determined retrospectively. Yet, in acting, we are also responsible for the consequences that our decisions have for distant and unseen stakeholders. To limit the notion of responsibility to only account for those who are well-represented in our conceptual schema, is to ignore all those who fall to the margins of the dominant discourse. This includes those disenfranchised stakeholders who do not necessarily have the economic clout or the social support systems to make their voices heard. As such, the over-determinations (i.e. the tensions and the competing interests) with which we have to grapple in order to be responsible, calls for interpretation and imagination. A measure of undecidability is essential for ethical decision-making. However, undecidability can never be used as an argument for inaction: we must act, even though our actions are imperfect. This means that when we act, we should exercise self-awareness, vigilance, and innovativeness, but we should also be willing to accept, trust, and engage in the viewpoints of others.

The justification for a complex, Derridean notion of corporate social responsibility is therefore based on the desire to mimimise the violence of our positions, and thereby the desire to remain open to otherness and to the future. Although economic imperatives are important, so too are social, environmental, and stakeholder issues; and, our current frameworks of thinking are proving inadequate for addressing the myriad problems that characterise these dimensions. It is therefore only in rethinking the nature of our responsibilities that we can begin to address these questions, and to reflect on the complexities that characterise our embedded positions in the world. Practically-speaking, we need to develop new conceptions of CSR and – having justified the case for a complex deconstructive notion of CSR in this chapter – an attempt is made to give content to such a conception of CSR in the following chapter.

References

Adorno, T. 1973. *Negative dialectics*, trans. E.B. Ashton. London: Routledge.

Anschen, M. 1970. Changing the social contract: A role for business. *Columbia Journal of World Business* 5(6): 8–14.

Aristotle. 1981. *The politics*, revised ed., ed. T.J. Suanders and trans. T.A. Sinclair. London: Penguin.

Ashley, C. 2009. Background note: Harnessing core business for development impact. *Overseas Development Institute,* Feb 2009.

Bataille, G. 1947. *Méthode de méditation in Oeuvres complètes*. Paris: Gallimard.

Bowen, H. 1953. *Social responsibilities of the businessman*. New York: Harper.

Caputo, J.D. (ed.). 1997. *Deconstruction in a nutshell: A conversation with Jacques Derrida*. New York: Fordham University Press.

Critchley, S. 1999. *Ethics-politics-subjectivity: Essays on Derrida, Levinas and contemporary French thought*. London/New York: Verso.

Dahlsrud, A. 2006. How corporate responsibility is defined: An analysis of 37 definitions. *Corporate Social Responsibility and Environmental Management* 15: 1–13.

De George, R.T. 2008. An American perspective on corporate social responsibility and the tenuous relevance of Jacques Derrida. *Business Ethics: A European Review* 17(1): 74–86.

Detrida, J. 1976. *Of grammatology*, trans. G. Spivak. Baltimore: John Hopkins University Press.

Derrida, J. 1978. From restricted to general economy: A Hegelianism without reserve. In *Writing and difference*, trans. A. Bass, 251–277. Chicago: University of Chicago Press.

Derrida, J. 1988. Afterword. In *Limited Inc*, ed. G. Graff and trans. S. Weber, 111–160. Evanston: Northern Western University Press.

Derrida, J. 1992a. 'Counterfeit money' II: Gift & countergift, excuse & forgiveness. In *Given time: I. counterfeit money*, trans. P. Kamuf, 108–173. Chicago: University of Chicago Press.

Derrida, J. 1992b. Force of law: The "mystical foundation of authority", trans. M. Quaintance. In *Deconstruction and the possibility of justice*, ed. D. Cornell, M. Rosenfeld, and D.G. Carlson, 3–67. New York: Routledge.

Derrida, J. 1994. *Specters of Marx: the state of the debt, the work of mourning & the new international*, trans. P. Kamuf. London/New York: Routledge.

Derrida, J. 1999a. Hospitality, justice and responsibility: A dialogue with Jacques Derrida. In *Questioning ethics: Contemporary debates in philosophy*, ed. R. Kearney and M. Dooley, 65–83. London/New York: Routledge.

Derrida, J. 1999b. *Sur Parole*. Paris: Editions de l'Aube.

Derrida, J. 2002a. On the "priceless," or the "going rate" of the transaction. In *Negotiations: interventions and interviews, 1971–2001*, ed. and trans. E. Rottenberg, 315–328. Stanford: Stanford University Press.

Derrida, J. 2002b. The right to philosophy from a cosmopolitical point of view: The example of an international institution. In *Ethics, institutions, and the right to philosophy*, ed. and trans. P.P. Trifonas, 1–18. Maryland: Rowman & Littlefield Publishers.

Derrida, J. 2002c. As if it were possible, "within such limits".... In *Negotiations: Interventions and interviews, 1971–2001*, ed. and trans. E. Rottenberg, 343–370. Stanford: Stanford University Press.

Derrida, J. 2002d. Negotiations. In *Negotiations: Interventions and interviews, 1971–2001*, ed. and trans. E. Rottenberg, 11–40. Stanford: Stanford University Press.

Derrida, J. 2002e. Ethics and politics today. In *Negotiations: Interventions and interviews, 1971–2001*, ed. and trans. E. Rottenberg, 295–314. Stanford: Stanford University Press.

Donaldson, T. 1993. Constructing a social contract for business. In *Business ethics: A philosophical reader*, ed. T.I. White, 167–187. New York: MacMillan Publishing Company.

Egels-Zandén, N., and M. Kallifatides. 2009. The UN global compact and the enlightenment tradition: A rural electrification project under the aegis of the UN global compact. *Corporate Social Responsibility and Environment Management* 16: 264–277.

Elkington, J. 1999. *Cannibals with forks: The triple bottom line of 21st century business*. Minnesota: Capstone.

Freeman, R.E. 2008. Managing for stakeholders. In *Ethical issues in business: A philosophical approach*, ed. T. Donaldson and P.H. Werhane, 39–53. Upper Saddle River: Pearson/Prentice Hall

Friedman, M. 1962. *Capitalism and freedom*. Chicago: University of Chicago Press.

Friedman, M. 1970. The social responsibility of business is to increase its profits. *New York Time Magazine*, 13 Sept 1970, 32–33 and 122–126.

Grosz, E. 2000. The time of violence: Deconstruction and value. *Cultural Values* 2(2/3): 190–205.

Hegel, G.W.F. 1977. *The phenomenology of spirit*, trans. A.V. Miller. Oxford: Clarendon Press.

Howells, C. 1998. *Derrida: Deconstruction from phenomenology to ethics*. Cambridge: Polity Press.

Jones, C. 2007. Friedman with Derrida. *Business and Society Review* 112(4): 511–532.

Jones, C., M. Parker, and R. ten Bos. 2005. *For business ethics*. Oxon: Routledge.

Kojève, A. 1969. *Introduction to the reading of Hegel: Lectures on the phenomenology of spirit*, ed. A. Bloom and trans. J.H. Nichols, Jr. New York/London: Basic Books.

Lee, M.D.P. 2008. A review of the theories of corporate social responsibility: Its evolutionary path and the road ahead. *International Journal of Management Reviews* 10(1): 53–73.

Levinas, E. 1974. *Autrement qu'être ou au-delà de l'essence*. The Hague: Nijhoff.

Levinas, E. 1990. *Totalité et infini*. Paris: Livre de Poche/Levinas, E. 1969. *Totality and infinity*, trans. A. Lingis. Pittsburgh: Duquesne University Press.

Matten, D., and J. Moon. 2008. "Implicit" and "explicit" CSR: A conceptual framework for a comparative understanding of corporate social responsibility. *Academy of Management Review* 33(2): 404–424.

Morin, E. 2007. Restricted complexity, general complexity, trans. C. Gershenson. In *Worldviews, science and us: Philosophy and complexity,* ed. C. Gershenson, D. Aerts, and B. Edmonds, 5–29. Singapore: World Scientific.

Morin, E. 2008. *On complexity*, trans. S.M. Kelly. Cresskill: Hampton Press.

Power, C. 2008. Manufacturing: The burden of good intentions. *TIME,* Wednesday, 11 June 2008. Available online at: http://www.time.com/time/magazine/article/0,9171,1813511,00.html.

Preiser, R., and P. Cilliers. 2010. Unpacking the ethics of complexity: Concluding reflections. In *Complexity, difference and identity*, ed. P. Cilliers and R. Preiser, 265–287. London/New York: Springer.

Raffoul, F. 2008. Derrida and the ethics of the impossible. *Research in Phenomenology* 38: 270–290.

Rorty, R. 1989. *Contingency, irony, solidarity*. Cambridge: Cambridge University Press.

Shaw, W. 2008. *Business ethics*, 6th ed. Belmont: Wadsworth Publishing.

Siponen, M. 2004. A pragmatic evaluation of the theory of information ethics. *Ethics and Information Technology* 6(4): 279–290.

Spivak, G.C. 1994. Responsibility. *Boundary 2* 21(3): 19–64.

Teubner, G. 2001. Economics of gift – Positivity of justice: The mutual paranoia of Jacques Derrida and Niklas Luhmann. *Theory, Culture & Society* 18(1): 29–47.

Wallich, H.C., and J.J. McGown. 1970. Stockholder interest and the corporation's role in social policy. In *A new rationale for corporate social policy*, ed. W.J. Baumol. New York: Committee for Economic Development.

Wood, D. 1999. The experience of the ethical. In *Questioning ethics: Contemporary debates in philosophy*, ed. R. Kearney and M. Dooley, 105–119. London/New York: Routledge.

Chapter 6
Towards a Theory and Model of Corporate Social Responsibility and Implications for Management and Leadership Practices

Abstract The deconstructive and complexity-inspired reading given of CSR in Chap. 5 is developed into a theory and model of CSR in this chapter. To this end, a critical investigation of the three components of CSR is presented, namely corporate identity, the nature of the relations between corporations and society (including stakeholders), and the nature of corporate responsibility. The model of CSR that is derived from this investigation depicts the different domains of CSR (namely, the environmental, social, legal, and economic domains) as embedded in one another, and differentially related to one another. It is argued that these domains are interlinked in complex ways with the corporation, which makes it impossible to conclusively define corporate responsibilities. These complexities also frustrate our attempts at managing our CSR obligations, as our current analytic tools are inadequate in dealing with these complexities. A number of management tools that can help in this regard are therefore also introduced, as is a discussion on the leadership approach and responsibilities that support the theory and model of CSR developed here.

Introduction

Gerald Midgley (2003: 93) argues that complexity thinking (of which deconstruction can serve as an example) leads to 'new and different questions about what forms of intervention we should pursue'. The analysis and defense of a complex, deconstructive notion of CSR undertaken in the previous chapter certainly poses many questions for our understanding of CSR. Yet, in order for these ideas to gain traction, it is important to try and flesh out the implications that this analysis holds for a theory and a model of CSR.

M. Woermann, *On the (Im)Possibility of Business Ethics*, Issues in Business Ethics 37,
DOI 10.1007/978-94-007-5131-6_6, © Springer Science+Business Media Dordrecht 2013

Perhaps a first skeptical question that comes to mind concerns the institutionalisation of these insights, namely: can we translate these insights into a substantive theory capable of informing and guiding our practices? In considering this question, let us first turn to Morin's (2008: 97) definition of complex thinking:

> [C]omplex thinking is not omniscient thinking. It is, on the contrary, a thinking which knows that it is always local, situated in a given time and place. Neither is it a complete thinking, for it knows in advance that there is always uncertainty.

Despite the uncertainties and contingencies introduced by any serious engagement with complexity, both Morin and Derrida are – as we have seen – of the opinion that the complexities that characterise our lives cannot excuse us from taking action. In Derrida's terms, we must assume a position, and this position must – moreover – be transgressive. We must explore beyond our current epistemologies and theories, and recognise that 'complexity does not put us only in the distress of the uncertain' but 'allows us to see besides the probable' and 'to explore the field of possibilities, without restricting it with what is formally probable' (Morin 2007: 29).

We can no longer afford to factor out complexity considerations from our theories and practices, including our business theories and practices. In a recently published Harvard Review article on managing complexity, Gökce Sargut and Rita McGarth (2011) argue that the central difference between today's business world and the business world of 30 years ago, is the level of complexity that people have to cope with. Sargut and McGarth attribute this rise in complexity to the proliferation of information technology in the last couple of decades, which means that systems which were formerly seen as functioning as closed and self-contained, are now viewed as interconnected, interdependent, and hence more complex. How this complexity is managed plays a crucial role in determining business success.

The call for action in the face of complexity is however issued with a warning, which we would do well to recap at this point: the fact that the models that inform our practices cannot capture phenomena in their full complexity, means that no theory can ever be understood as a finished project. As such, the theory and model of CSR presented in this chapter, represents a first step in trying to do homage to the complexities with which we should grapple when thinking about our corporate social responsibilities, but are both open to further interpretation and deconstruction.

The theory of CSR proposed here is developed at the hand of an investigation of the three components of CSR, namely the identity of *corporations*, the nature of the relations between corporations and *society* (including corporate stakeholders), and the nature of corporate *responsibility*. These components are, of course, inter-related and context-dependent issues. Adding to the complexity is the fact that even within a single organisational context, it is impossible to exhaustively define corporate responsibilities due to the diversity and changing nature of social issues (Polonsky and Jevons 2006). However, as with all theories, this complexity-inspired theory of CSR constitutes a meta-position that can be used to guide CSR practices. Such a position highlights the type of considerations that should be kept in mind when engaging in the particularities of a situation from a complexity viewpoint.

This meta-position can be useful, given that the existing knowledge on how to navigate complexity 'hasn't permeated the thinking of most of today's executives or the business schools that teach tomorrow's managers' (Sargut and McGarth 2011: 70). Moreover, Sargut and McGarth note the inadequacy of many of the analytic tools that managers functioning in complex environments employ. In order to address this challenge, a number of tools that can aid business practitioners in exercising their corporate responsibilities in a complex world will also be introduced. The type of leadership approach that supports the proposed CSR theory developed in this chapter will also be investigated, as will the responsibilities that leaders have in ensuring that organisations remain open to the complexities that define their operating environments.

Corporate Identity and Responsibility

Corporate Moral Responsibility: A First Look

The Traditional View

Over the past three decades, a central question in the business ethics literature is whether the concept of moral responsibility can be extended to organisations and their actions. In other words, the question is: can organisations be morally responsible for their actions in the same way that individuals are? In Woermann (2010), the gist of the debate is described as follows: on the one hand, it is argued that organisational systems and processes make it possible for organisations to undertake intentional actions that surpass the actions of individual corporate agents (French 1979, 1984; Erskine 2003; Petit 2007). On the other hand, it is argued that organisations are incapable of undertaking moral obligations because they function like machines, and are therefore only able to pursue empirical objectives (Ladd 1970; Werhane 1980); or, in a related argument, that organisations are incapable of moral motives and actions, as only biological agents can be defined as intentional agents (Keeley 1981, 1988; Velasquez 1983). Although the definition of intentionality may vary slightly within this debate, moral agency is conceptualised in terms of the causal actions of willing and capable actors.

Challenges to the Traditional View

Our complex realities challenge this view of moral responsibility, and in this regard, Sargut and McGarth (2011) argue that, in their observations, the two central problems facing managers of complex systems are *unintended consequences* and *difficulties in sense-making*. With regard to the problem of *unintended consequences*, they provide both positive and negative examples in the business world of 'events that interact without anyone meaning them to' (71; italicised in the original). The significance of this discussion is that it sheds light on the fallacy of intentionality. In a simple world, it is easy to act intentionally, however the more complex the situation becomes, the less it is within our control. In a related article, Michael Mauboussin (2011) attributes the endurance of the belief in intentional, goal-directed action to a strong 'cause-effect' bias. When asked what prevents us from dealing effectively with complexity, he responds that:

> The biggest issue... is that humans are incredibly good at linking cause and effect – sometimes too good. Ten thousand years ago most cause and effect was pretty clear. And our brains evolved to deal with that (90).

The world today is vastly different from the world of 10,000 years ago, and increasingly, we *struggle to make sense* of complex situations. This is in part due to what Sargut and McGarth (2011) term the *vantage point problem*, which arises because we act from a specific position, which inhibits us for making sense of the larger network of relations in which we are embedded. Our sense-making abilities are further hampered by our *cognitive limitations*, which makes it difficult for us to understand the effects of our own and other people's actions. Indeed, Sargut and McGarth argue that research indicates that executives typically believe that they absorb, and make sense of, more information than is really the case. Other challenges to sense-making mentioned by Sargut and McGarth, concern our propensity to *inattentional blindness*, which means that we tend to ignore the environment when focusing on a specific task; as well as *rare events,* that is, events that occur so infrequently that we are unable to understand how they impact on the system. Collectively, these problems pose a huge challenge to our understanding of moral responsibility, and the traditional debate on whether organisations can be seen as moral persons thus proves a fruitless avenue to pursue further, as it fails to recognise that both moral agents and organisations are embedded in a complex network of relations.

However, apart from the problematic manner in which moral agency is conceptualised within this debate, a second problem concerns the fact that this debate is based on the mistaken premise that the identity conditions of individuals (including intentionality, autonomy, and rationality) are *a priori* givens (Woermann 2010). In Chap. 2, it was argued that, given the complexity view, the identity of individuals and organisations are coterminous (in that our practices lead to the emergence of institutions, and our institutions serve to constrain our practices through feedback loops), and that the focus of any ethical analysis should center on the *relations* between individuals and the systemic properties that emerge from these relations. As such, one of the driving questions in the CSR debate should be what the emergence of corporate identity implies for our understanding of corporate social responsibility.

Corporate Identity

Since identity is a fluid concept, organisational identity should be conceptualised in terms of 'the stability of the labels used by organizational members to express who or what they believe the organization to be' (Gioia et al. 2000: 64). Although these labels tend to be relatively durable, it is important that organisational members (especially leaders) engage in dialogue regarding the organisation's identity, and accept responsibility for the manner in which their actions influence the direction of 'their' organisations. This is a crucial task, because, as Kurt Richardson and Michael Lissack (2001: 36) note, '[a] sense of identity is important because it determines how an individual directs his or her attention', and the same applies for organisations.

Institutions of Integrity and the Integrity of Institutions

One should specifically acknowledge the impact that corporate actions have on what Eduard Grebe and Woermann (2011) refer to as the 'institutions of integrity' and the 'integrity of institutions'. 'Institutions of integrity' represent the structurally and contextually-determined norms and codes that 'bind' individual behaviour. These norms and codes are, in part, based on underlying normative principles, but also emerge from contextually-defined practices. The 'integrity of institutions', on the other hand, is understood as 'correct functioning' and fitness for purpose – which, in the case of organisations, refers to the organisation's ability to ethically and effectively achieve its goals within a larger operating environment. These two components are interdependent in that our contextually-defined practices determine whether the organisation displays fitness for purpose. In this regard, consider the example of Enron: management's 'win-at-all-costs' focus helped to create an egotistical organisational culture that pushed profits at the expense of the company's long-term survival, ultimately leading to the company's ruin (see Sims and Brinkmann 2003). In this example, the institutions of integrity served to undermine the integrity of the institution.

Challenges to a Healthy Corporate Identity

In order to promote a robust organisational identity that can serve as 'a rudder for navigating difficult waters' (Albert et al. 2000: 13), it is important to continuously remain aware of both the context (or network of relationships) in which an organisation is embedded, and the direction of 'organisational becoming'. A healthy corporate identity is undermined by the *vantage point problem*, where corporate members lose sight of the impact that corporate actions have on stakeholders and the larger operating environment; *inattentional blindness*, which results when corporations or departments act as discrete units that narrowly focus attention on immediate corporate goals, thereby undermining corporate productivity; and *cognitive limitations*, which is exacerbated in corporations lacking in cognitive diversity. One way in which to counteract these challenges is through promoting organisational openness, because the capacity for dealing with complexity cannot exist in a fully constrained system (Cilliers 2010).

Achieving Organisational Congruence

When organisational practices or norms are incongruent with the operating environment, or when practices and norms lead to destructive conflicts, it is – as Jane Collier and Rafael Esteban (1999) warn – a sure sign to start planning as 'open systems' in order to survive. In Woermann (2010), it is argued that planning as open systems does not present a once-off organisational intervention, but is a process of continual re-organisation in an effort to create greater flexibility, and organisational robustness or resilience, where resilience is defined as an organisation's capacity to deal with stress and promote 'learning, self-organization, and adaptation at multiple scales' (Lissack and Letiche 2002: 82). Michael Lissack and Hugo Letiche further argue that an organisation's capacity for resilience is determined by the structure of the system and the interactions between the components of a system, as well as by organisational perceptions regarding change (especially unexpected emergent events). From this description, one can infer that more 'loosely coupled' organisational structures are better able to deal with current concerns such as globalisation, increased communication possibilities, technological change, financial innovation, freer trade possibilities, and heightened competition for market share (Collier and Esteban 1999).

Opening up work practices and creating greater organisational flexibility demands both that institutions of integrity are aligned with the integrity of the institution, and that the organisation (as an institution) is responsive to environmental complexities and contingencies. In other words, it is important to establish a sense of congruence within the organisation, as well as between the organisation and the environment, where congruence is understood as 'the ability to accommodate difference and dissensus, without losing [the system's] functional unity or sense of purpose' (Painter-Morland 2008: 224).

Corporations are more likely to achieve congruence if corporate members are morally aware. As will be elaborated upon in the penultimate section of this chapter, moral awareness is specifically critical for planning as open systems, avoiding blame-shifting and apathy, promoting an extended view of our actions, and nurturing an understanding of the nature of our models and strategies (including models and strategies pertaining to CSR). Being morally aware and responsible also necessitates that we take cognisance of the fact that organisational congruence may be temporary, and that our decisions and actions are subject to risks, hazards, derailments, and transformations (Morin 2008) – all of which impact upon the integrity of our institutions. In this regard, moral responsibility implies facing up to uncertainty, dealing with matters beyond the present (even if our long term perspectives are inherently flawed), and being flexible and responsive enough to try and correct decisions and actions with undesirable consequences.

In showing how our actions impact on 'institutions of integrity', and the 'integrity of institutions' (and thus organisational identity), the above analysis helps to dispel the myth that being ethical is easy, since such an idea ignores the impact that our work practices and identities have on our views of our moral responsibilities. Furthermore, the analysis also serves as a challenge to the commonly accepted view that unethical behaviour in business is simply the result of 'bad apples' (Trevino and Brown 2004). Although Linda Trevino and Michael Brown concede that there are bad actors, they argue that 'most people are the product of the context they find themselves in. They tend to "look up and look around," and they do what others around them do or expect them to do' (72). In this regard, CSR starts at home, since how we define our social responsibilities is contingent on our definition of the corporation, and of ourselves as corporate members.

Conceptualising Social Responsibility

The preceding discussions on corporate moral responsibility and corporate identity set the stage for developing a complexity-inspired model of corporate social responsibility. Before presenting this model, a critique of Archie Carroll's (1991) model of CSR, as well as of Mark Schwartz and Archie Carroll's (2003) subsequent refinement of this model will be offered. Both these models have not only been influential in framing CSR (indeed, Schwartz and Carroll (504) report that 'Carroll's CSR domains and pyramid framework remain a leading paradigm of CSR in the social issues in management field'), but are also compatible with the traditional view of CSR as articulating an equalising justice.

Traditional Models of CSR

The Pyramid of Corporate Social Responsibility

In his influential article, entitled 'The pyramid of corporate social responsibility: towards the moral management of organizational stakeholders', Archie Carroll (1991) identifies four types of social responsibilities constituting CSR, namely: economic, legal, ethical and philanthropic responsibilities. He represents these dimensions of CSR in terms of a pyramid (see Fig. 6.1), where the lower rungs, namely the economic and the legal, are required by organisations, in that organisations must operate profitably and legally in order to be sustainable; the middle rung, namely ethics, is expected of corporations; and the top rung of the pyramid, which constitutes philanthropic responsibilities, is deemed desirable as it promotes good corporate citizenship.

Although this CSR pyramid has been employed by numerous theorists and empirical researchers, Schwartz and Carroll (2003) note several problems with this schema. Firstly, they argue that the depiction of corporate responsibilities in terms of a pyramid suggests a hierarchy of CSR dimensions, with philanthropic responsibilities being the most important. Although Carroll explicitly states that the economic and the legal dimensions are the most important, the use of the pyramid to illustrate CSR responsibilities can nevertheless cause confusion.

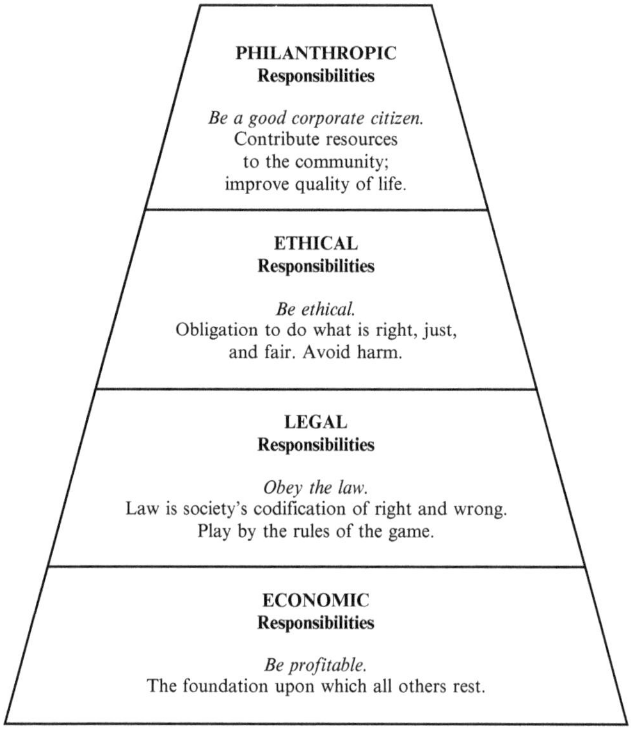

Fig. 6.1 The pyramid of corporate social responsibility (Carroll 1991: 42)

Secondly, the pyramid framework cannot accommodate overlapping dimensions of corporate responsibility. Thirdly, the incorporation of the philanthropic dimension causes confusion, as philanthropic activities are voluntary and therefore do not strictly speaking fall under the domain of CSR; and, where they arguably do, they can readily be accommodated in the ethical domain (although this causes further confusion as to what ethical responsibility entails). Lastly, Carroll does not provide much discussion on how corporations can engage in multiple domains, and nor are his discussions of the four domain complete.

The Three Domain Model of Corporate Social Responsibility

For these reasons, Schwartz and Carroll (2003) propose a refined, three domain model of CSR (see Fig. 6.2), which is depicted as a Venn diagram, and which can accommodate seven categories of corporate responsibilities (four of which constitute an overlap of responsibilities). This new depiction expands on the ways in which CSR can be conceptualised, analysed, and illustrated, and is especially useful in highlighting the forces that come into play in ethical decision-making.

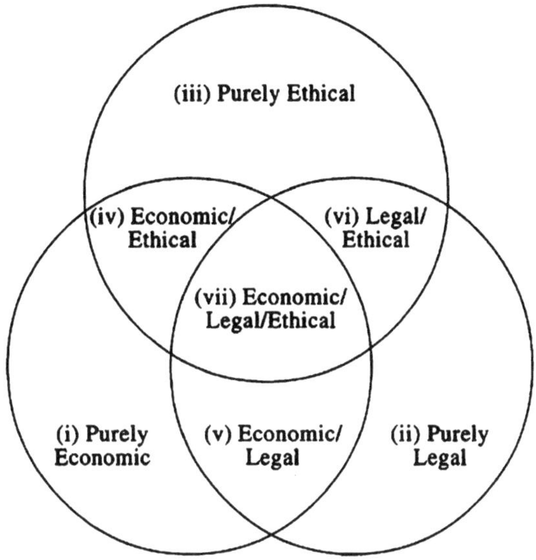

Fig. 6.2 The three domain model of corporate social responsibility (Schwartz and Carroll 2003: 509)

Despite the superiority of this model over the initial framework, Schwartz and Carroll also note a number of limitations of the new model, most of which are linked to the assumptions in which the model is grounded. Schwartz and Carroll (520) elaborate on these assumptions in the following passage:

> The model assumes that the three domains of CSR are somewhat distinct, and that they are all-encompassing. In terms of being somewhat distinct, some might question whether any action can be defined as "purely economic," "purely legal," or "purely ethical." In other words, one might argue that economic, legal, and ethical systems are all interwoven and inseparable… In terms of the three-domain model being all encompassing, it is not clear whether there are corporate activities which are engaged in without reference to at least their economic impact, the legal system or ethical principles.

From the perspective of this study, these are devastating limitations, since the assumptions regarding the distinctive and all-encompassing nature of CSR categories amounts to a rejection of complexity. This is because the domains of CSR are not treated in relational terms, but in terms of the paradigm of disjunction (in which – to recall – cognitive difficulties are separated from one another), and the paradigm of universalism (where CSR categories are exhaustively defined in terms of economic, legal, or ethical duties).

A further problem related to the distinctive nature of the CSR categories concerns trade-offs. This problem is discussed in Johan Hattingh's (2006) critique of standard depictions of sustainability models, and the same argument can be made in this context. Despite drawing attention to the interdependence between the ethical, legal, and economic realms, the above representation of CSR reinforces (rather than undermines) the notion that the three central categories of CSR are external to one another and function according to their own rules. Given this interpretation, managing one's CSR obligations amounts to finding the right

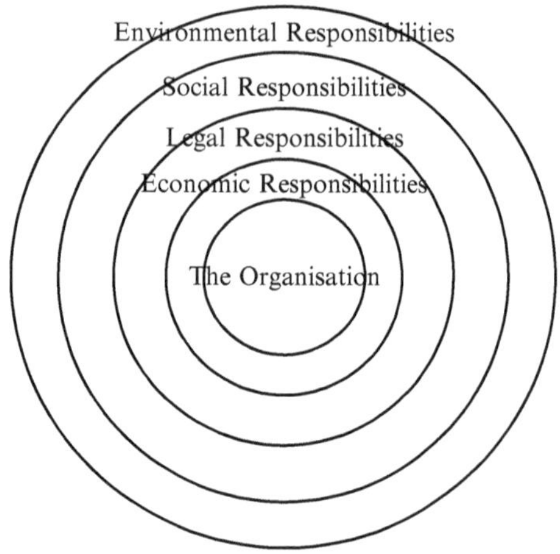

Fig. 6.3 The embedded model of corporate social responsibility

balance or optimal trade-off between these categories. However, determining the correct balance often ends up being a function of the set of values that is given priority by management, which in most cases will amount to financial responsibilities. In practice, this implies that a corporation's social responsibilities can too easily be interpreted as 'nice-to-have', rather than integral to business itself.

An Alternative Model: The Embedded Model of CSR

Given the above critique, it is posited that a complexity-inspired model of CSR, like a complexity-inspired model of sustainability, 'would be one in which these... spheres were conceptualised as *embedded* within one another, with values that are internally linked to one another and a logic that inseparably intertwines them' (Hattingh 2006; 204; my italics). However, not only would a complexity-inspired model be based on an alternative diagrammatic representation of CSR, but the very categories of CSR would also differ from Carroll's and Carroll and Schwartz's models. Specifically, it is proposed that if every business decision has an ethical dimension (the ethics of complexity), and if ethics cannot be determined in advance (the complexity of ethics), but results from the complex relations between corporate stakeholders and the larger environment, then ethics should be viewed as an emergent property – one that cannot be captured within a CSR model, but that nevertheless informs the model. Controversially then, a model of CSR without explicit reference to an ethical domain is advanced. This model (see Fig. 6.3) serves to illustrate the embedded nature of the organisation within the economic, social, legal, and environmental spheres (each of which generate organisational responsibilities), and serves to draw attention to the networks in which the organisation partakes, and which help to constitute an organsisation's identity. Again, the relations between these spheres should be understood differentially as a network of relations framed in a specific manner, and not in terms of binary oppositions, as has been the case in traditional depictions of CSR. In other words, the dynamics between these spheres are best described in terms of the notion of *différance*, which leads to a fluid characterisation of systems identity, premised on notions of difference and deferral. This complex view of a system's identity also precludes a simple identification and formulation of responsible action, thereby making it impossible to pinpoint something like a 'purely ethical' sphere.

Differentially-Related Domains

Despite the complexities involved, some analytic observations can nevertheless help to elucidate the nature of this model. Specifically, this model shows how an organisation is differentially-defined from a natural environment, without which it could not exist. This seems like a trivial point, but in drawing attention to the natural environment and its relations with the other spheres, managers are more likely to realise that, over the long-term, natural resources cannot be 'traded-in' for some human benefit without consequence, since the very definition of human benefit is already contingent on a certain understanding of the natural world. Organisations therefore have a responsibility to operate in an environmentally-sustainable fashion. Secondly, the organisation is embedded in a social environment, which is in itself understood in relation to this natural environment, and which serves to remind us that organisations exist primarily to provide goods and services to society, and should therefore be run to the benefit of their stakeholders, which are critical to any organsiation's survival. Thirdly, the organisation exists in a legal environment, and the legal environment determines the economic 'rules of the game', as recognized by Friedman (1962) and Smith (1776). Therefore, organisations have a responsibility to generate profits for shareholders without recourse to deception, force, or fraud. Both the economic and legal realms are however situated in the wider social and natural realms, which reinforces the point that the larger operating context matters, and that an organisation's identity and responsibilities cannot be determined outside of context.

Since these various domains are differentially-related, responsible corporate actions should also be informed by an in-depth engagement on how the different domains co-constitute one another and how corporate actions impact on all levels. For example, in order to responsibly generate profit for shareholders, corporations must account for the impact that their actions have on the environment, their treatment of other corporate stakeholders, and the rules and laws that restrict their activities. Again, this might seem self-evident, but the recent financial crisis attests to the myopic vision of many large and influential institutions. The proposed model of CSR also reinforces the importance of organisational openness: it is only by remaining open to the environment that an organisation is able to nurture an organisational identity that allows it to function in a sustainable and ethical manner. The environment thus becomes an integral part of the organisations activities, and the boundary between the organisation and the environment acts as an interface that participates in constituting both the organsiation and its environment. This implies that the focus of ethical evaluation should shift from the perceived core of the organisation to its periphery (Collier and Esteban 1999; Cilliers 2001).

Although certain domains depicted in the model will receive priority in certain situations, the other domains cannot be ignored. In this regard, the metaphor of the 'Borromean knot', in which three rings are interlinked in such a way that one cannot remove a link without destroying the whole, proves useful. How the model is interpreted is also a function of the level of analysis employed. In other words, the domains can be viewed in a local or a global light, which alters the type of considerations that define each domain, and the concurrent corporate responsibilities that arise from an engagement with these domains. Moreover – and as already mentioned above – given the complex ways in which these domains impact on one another and on the organisation, it is impossible to exhaustively define our responsibilities, and – if necessary – additional domains can easily be added to tailor the model to an organisation's specific context, and to extend the scope of corporate responsibility.

Overcoming the Challenges to Corporate Social Responsibility

Unlike traditional models of CSR, the proposed model (which views corporate identity in terms of differences) draws attention to the importance of diversity, as a means of developing a rich and robust corporate identity. Allen (2001) distinguishes between 'requisite diversity', which denotes the minimal level of variety needed for a system to cope with its environment, and 'excess diversity', which allows systems to experiment internally, and thereby generate a number of strategies for operating in a given environment. He further argues that excess diversity is needed for long-term systems' survival, since the 'fat' of excess knowledge and diversity is necessary for experimenting and innovating for the future. Diversity is therefore critical for countering the consequences that arise from our *cognitive limitations*, in that it leads to internal experimentation and the generation of a variety of strategies for coping with the operating environment. Indeed, both Sargut and Gunther (2011) and Mauboussin (2011) raise this point in their discussions on managing complex systems, respectively arguing that: 'you need to make sure your organization contains enough diverse thinkers to deal with the changes and variations that will inevitably occur' (76) and '[c]ognitive diversity – intentionally putting together different points of view that will challenge one another – is essential for hiring and building teams' (91). Furthermore, drawing attention to the embedded and differential nature of corporate identity helps to counteract the *vantage point problem* as the focus lies on the boundary between the organisation and its environment, and the reciprocal influences that the environment and the organisation exert on each other. In terms of this model, we are also less likely to succumb to *inattentional blindness*, since corporate tasks should be framed within a larger organisational and environmental context.

Extending the Scope of CSR

In Chap. 5 it was argued that what is lacking in traditional conceptions of CSR is a *critical* reflection on how our embedded practices shape our views on morality and responsibility (as enacted in CSR practices). In this regard, the proposed model of CSR can help to initiate such a critical discussion by drawing attention to the complex network of relations in which a given business or industry is embedded. However, in order to truly act responsibly, it is also important for business managers and leaders to engage with the limitations of our models and to acknowledge that we have incomplete knowledge regarding the impact of our actions. This will also help to guard against an equalising justice, where business responsibilities are merely articulated in terms of a well-understood social contract. This is because taking account of the complexities of our operating environments, and of our own finite resources for dealing with these complexities, raises our awareness of the fact that we cannot conclusively determine our responsibilities. Rather, responsible action is the product of the carefully-considered decisions that we make when faced with the complexity and uncertainty of 'the market economy' which 'is vast, and unknown, resulting in 'consequences of the incalculable'' (Derrida 2002: 321). Responsible action requires that we push against the limits of our understanding, in attempting to answer the call of a justice which 'cannot be reappropriated' (Derrida 1997: 18), and which therefore exceeds the finite circle of the *oikos* or closed economy and the calculable, symmetrical structure of an equalising justice. In this section, some examples of the uncertainties that define the environmental, social, economic, and legal domains are discussed, in order to highlight the extended scope of business responsibilities within a complexity-inspired model of CSR. However, it must be noted that within this model, identifying corporate responsibilities remains a context-dependent activity, and it remains for each organisation to reflect on, and define, their unique corporate responsibilities.

Environmental Responsibilities and the Broader Sustainability Challenge

At a basic level, sustainability is concerned with system's maintenance, which means that our actions should not impact on a system in ways that threaten its long-term viability (Crane and Matten 2004a). Since all human institutions (including the organisation) are embedded in the natural environment, it stands to reason that – given the complexity-inspired CSR model presented above – environmental sustainability, which concerns 'the effective management of physical resources so that they are conserved for the future' (24), warrants critical attention. Andrew Crane and Dirk Matten argue that biosystems and natural resources are finite, and are negatively impacted upon by a number of human activities, including industrialisation, the continued use of non-renewable resources, and the use of damaging environmental pollutants. Yet, determining the extent or nature of these threats on the environment is very difficult, which in turn means that we are unsure of what constitutes sustainable practices. Questions with which we should grapple include: 'What are the time-scales involved? Are there measures for sustainability and, if so, to whose benefit are they applied? Will what seems to be a sustainable strategy now actually prove to be so in the future?' (Cilliers 2008: 40).

The Challenge of Globalisation and the Role of Business

The difficulties inherent to thinking about sustainability are further compounded by globalisation, which necessitates that our actions be evaluated on a planetary scale. Crane and Matten (2004a: 362) argue that 'globalization increases the demand for business ethics because deterritorialized or transnational spaces such as global financial markets or the ozone layer are beyond the control of national (territorial) governments'. However, globalisation also challenges our understanding and practice of business ethics, because the complexity generated by countless bits of information about the world means that 'the more we are grasped by the world the more difficult it is for us to grasp it' (Morin 1999: 31). Morin argues that it is precisely this complexity that makes us long for the possibility of isolating a vital problem that would help us to resolve all our other problems. Unfortunately, no one vital problem exists, because, as Morin (31) explains, '[t]he planetary problem is a whole fed by multiple, conflictual, crisical [*crisique*] ingredients; it encompasses, surpasses, and feeds them in return.' It is therefore important to account for the complexities that define our planetary problem, whilst simultaneously taking cognisance of the fact that despite these complexities, we are all 'connected in the same planetary community, sharing a common fate' (38).

Contrary to the assumptions underscoring the separation thesis (see Chap. 5), business decisions cannot be separated from business responsibilities, and – because of its immense reach, power, and influence – business should become a central actor in the sustainability movement. Furthermore, business responsibilities should not be articulated in semantics that calls for 'sacrifices' or the 'taming of capitalism' (as has often been the case in the governance literature). Such language frames companies as opponents, rather than as partners in working towards a sustainable future (Homann 2007). As Homann (7) states: 'Corporations cannot be expected to make 'sacrifices,' but they can *invest* – in real or human capital, but also in the social order as a prerequisite of long-run benefit.'

Rethinking Development and the Role of Business

If we want the sustainability movement to be successful, we need to rethink our understanding of development. All-too-often, development is conceived exclusively as techno-economic progress, which mostly has negative effects on the natural environment. In order to work towards sustainable strategies, Morin (1999: 34) argues that we need 'a more rich and complex notion of development', and we should understand development as 'joint development of individual autonomies, community participations, and a sense of belonging to the human species' (25). Playing in the park 'with big business and grand politics' (Cilliers 2008: 54), or blindly endorsing a scientific rationality will not help in solving our problems.

The idea of joint development also implies a rejection of the paradigm of disjunction, in favour of a complex understanding of cause and effect. Joint development requires vigilant thought and careful action, and in this context, it is helpful to recall Derrida's (1999: 68) view of Hamlet:

> if we assume that Hamlet is a figure of paralysis or neurosis because of undecidability, he might also be a paradigm for action: he understands what actions should be and he undergoes the process of undecidability at the beginning.

Techno-optimists, who refuse to acknowledge the complexities that characterise our planetary problem, tend to avert their eyes when faced with the unnameable prospects that our actions hold for the future. Responsible thought and actions require that we face 'the terrible process of undecidability' (66), and, thereby, acknowledge that sustainable action cannot be articulated in a clear-cut programme, but requires transgressive and imaginative thinking. Environmentally sustainable practices, like sustainable social and economic practices, necessitate that we move beyond accounting for the negative externalities that corporate actions have on the natural and operating environments. Business as a whole should also accept responsibility for its role in securing and improving upon our planetary future, since, as Homann (2007) argues, corporate governance will only be successful if companies bear the responsibility for governance collectively. Currently, there is still little coordination between companies; and, according to Homann (5), a key reason for this is that corporations are not 'used to cooperating with other corporations in the area of global politics – in mutual interests – while remaining rivals in markets.' Therefore, it is only in focusing on joint development and our common human future that it becomes possible to change the mindset of companies, who still see themselves as operating in isolation to one another, and in isolation from the human community.

Social Responsibilities: Accounting for Stakeholders

Although the concept of sustainability has its roots in environmental management and analysis, the concept has been extended to include social and economic aspects, as is clear from the above discussion. One reason for this is that, since the sustainability movement is concerned with intergenerational welfare, it should, logically, also be concerned with the equity of current generations (Crane and Matten 2004a). In this regard, corporations have a large role to play in accounting for the effects of corporate actions on stakeholder equity.

Complex Relations

Given a complexity perspective of identity formation, it is sensible to follow Painter-Morland (2008) in arguing that we are accountable *towards* those who co-constitute the network of relations to which we belong, as opposed to merely being accountable *for* the causal effects of our actions. The former represents a legal understanding of business responsibility, which amounts to a form of blame responsibility; whereas the latter represents an obligation responsibility (in that, to fail to act appropriately is to fail *someone*). In this regard, Painter-Morland draws on the etymology of the word 'responsibility', arguing that in Latin, *respondre* means to answer, or to promise in return (pledge back) to those with whom one associates.

From a complexity perspective, it is also important to consider what 'association' with others implies. As argued in Chap. 5, the primary focus of standard ethical theories is 'on the agent committing the actions, not the object (or patient) receiving those actions' (Siponen 2004: 281). These theories are therefore anthropocentric, logocentric, and egocentric (Floridi 1999: 42). Translated into organisational terms, this would imply an emphasis on the organisation or company, rather than on the stakeholders (who are affected and influenced by the organisation's actions). Such a view creates the impression that organisations occupy the central position on stakeholder maps, which implies both that the organisation interacts freely with stakeholders on its own terms, and that it operates as a self-contained entity (Painter-Morland 2006).

If we take a complexity view of identity formation seriously, then critical questions regarding the logical and linear demarcation of agent-patient remain unsolvable – to recall the words of Derrida (1995b: 285), 'we never know, and never have known, how to *cut up* a subject.' Understanding the ways in which we are interconnected with one another necessitates complexity thinking, which in turn requires not only a move beyond agent-centric ethics, but also a move beyond biocentric and patient-orientated ethics, which 'is centred on, and interested in, the entity itself that received the action, rather than its relation to or relevance to the agent' (Floridi 1999: 42). From the perspective of a complexity-centric ethics, one must think the inextricable *relation* between self and other, whilst simultaneously respecting the other's difference – not only in terms of manifest differences, but also in terms of differences that cannot be thought within our conceptual models.

Rethinking Stakeholder Theory

Applied to CSR, one can argue that stakeholder theorists, such as Edward Freeman, have done significant work in drawing attention to the groups (beyond shareholders) that affect and are affected by the organisation's actions. In Freeman's model, responsibilities to stakeholders are not formulated in terms of abstract, general imperatives, but exist between discrete entities (Painter-Morland 2006). Freeman and Philips (2002) view stakeholder relations in terms of value-creating activities, which are defined as voluntary agreements that are the product of co-operating free wills. Although Freeman (2008) draws a distinction between primary and secondary stakeholders, all stakeholder relations are analysed in terms of the reciprocal obligations that stakeholder groups and the organisation have towards one another. As argued in Woermann (2011), conceptualising such relationships requires sensitivity to the contingencies that may impact on the nature of the relationships, and it is in this regard that the managerial task is defined not simply as describing existing stakeholders or predicting cause-effect relationships, but also recommending attitudes (Donaldson and Preston 1995). In other words, the managerial task is not merely descriptive or instrumental, but also normative.

Carroll (1991: 43) argues that in this regard '[m]anagement's challenge is to decide which stakeholders merit and receive consideration in the decision-making process', and he further deems legitimacy and power to be two vital criteria for deciding on key stakeholders. However, for Carroll the criterion of power is only considered in terms of the influence that certain stakeholder groups wield over the organisation. What is missing from both Carroll's and Freeman's (and Philips') analyses is a consideration of the amount of discretionary power invested in the organisation's management, and the effects that this power has on the legitimacy and urgency of stakeholder interests.

Since power determines the ability of certain actors to bring about the outcomes that they desire (Salancik and Pfeffer 1974), it has a big effect on the perceived legitimacy and urgency of stakeholder claims. This is because stakeholders with legitimate claims on the organisation will not achieve salience for the organisation's management, unless they have the power to enforce their wills upon the organisation, or create the perception that their claims are urgent (Mitchell et al. 1997). For this reason, Woermann (2011) argues against Freeman and Philips' view that all manager-stakeholder relations are consensual, voluntary agreements.

Woermann (2011) further argues that ignoring the specific nature of the relations between the organisation and its various stakeholder groups could moreover lead to a situation where stakeholder relations that cannot easily be accommodated in Freeman's model are largely ignored. The fact that management is often vested with high levels of discretionary power, and the fact that many stakeholders find it difficult to articulate their claims on the organisation due to the complex and pervasive effects that corporations have on society, means that responsibility cannot be limited to merely accounting for well-represented and articulated stakeholder groups. Management should continuously reflect on ways in which to also account for disenfranchised, dispossessed, and distant stakeholders, who rarely feature in their thinking. In this regard, Badiou's (2009) tribute to Derrida (introduced in Chap. 4) is poignant: responsible action is necessarily transgressive. It constitutes the attempt to account for the *inexistant*, who exist beyond our current frames of meaning, but who nevertheless often bear the costs of our actions.

Furthermore, in attempting to manage stakeholder relations, managers should simultaneously consider the corporation's relation with a given stakeholder group, whilst acknowledging the claims of other stakeholder groups (who constitute the face of the third (Derrida 1999) that serves to complicate any singular relation with a given stakeholder group). As such, management should 'take account of interactions and interdependencies among the network of stakeholders, rather than focus on individual firm-stakeholder relations' (Crane and Matten 2004b: 366). Only by acknowledging these complex interdependencies, will we come to realise that the greatest difficulty in managing stakeholders is that there is 'too much determinacy' (Derrida 1999: 79). Contrary to Freeman's (2008: 48; bolded in original) argument that '[m]anaging for stakeholders is about creating as much value as possible for stakeholders, without resorting to tradeoffs', a complexity-informed understanding of stakeholder theory requires that we acknowledge the inevitability of trade-offs and accept responsibility for these trade-offs.

Economic and Legal Responsibilities: Defining and Accounting for Systemic Risks

Unlike the environmental and social domains, the economic and legal domains are guided by fairly rigid institutions, in which the appropriate rules of engagement are clearly articulated. Schwartz and Carroll (2003) argue that within the economic domain, business has the responsibility to maximise profits and/or maximise shareholder value through undertaking direct economic activities (that increase sales whilst avoiding litigation) and indirect economic activities (designed to heighten employee morale and the company's reputation). These economic activities are conducted within the broader 'rules of the game', which help to determine which activities are appropriate and which are not. However, although these economic duties are important, management should also recognise that rules and laws are fallible and that one cannot legislate for every situation.

Economic Responsibilities

Colander et al. (2009) argue in an article entitled 'The financial crisis and the systemic failure of the economics profession' that the economic models that have been developed over the past three decades, and on which the 'rules of the game' are based, 'disregard key factors – including the heterogeneity of decision rules, revisions of forecasting strategies, and changes in the social context – that drive outcomes in asset and other markets' (250). In other words, their contention is that these economic models do not successfully model the real world, since most models still endorse equilibrium conditions, which mean that markets are viewed as inherently stable, and market deviations are seen as merely temporary. Furthermore, they argue that the failure of economists to predict the recent financial crisis is partly due to the widely-held view that economics concerns the allocation of scarce resources, which 'reduces economics to the study of optimal decisions in well-specified choice problems' (251). As with the moral agent, the economic agent is construed as a free, intentional agent with perfect information; and, in these models, economics is restricted to the realm of the *oikos*, which, to recall, is governed by 'the limits of need, of the useful, of the natural, the reasonable, the calculable, the stable relation between production and consumption, between the *chez soi* and the *chez l'autre*' (Derrida 2002: 321). In failing to account for chrematistics or the an-economic (to use Derrida's terminology), Colander et al. (2009: 251) argue that:

> Such research generally loses track of the complex dynamics of economic systems and the instability that accompanies it. Without an adequate understanding of these processes, one is likely to miss the major factors that influence the economic spheres of society.

Morin (1999) concurs with this point, in arguing that, despite being extremely specialised (indeed, economics is the most mathematically advanced social science), economics remains a 'humanly backward science'. The reason for this is because economics 'has abstracted itself from the social, historical, political, psychological, and ecological conditions inseparable from economic activity' (17).

According to Colander et al. (2009), financial economists are morally culpable for the financial crisis, because they failed to warn the public about the abstracted-nature, and hence the fragility, of these models. 'Economists' they state '. . .have an ethical responsibility to communicate the limitations of their models and the potential misuse of their research' (252). However, market participants (including business) as well as policy-makers are also to blame, primarily for defining systemic risk outside their sphere of responsibility. In this regard, Colander et al. state that '[t]he neglect of systemic externalities by market participants and policy makers is not only unethical; it is a prudential lapse as well' (254). By narrowly defining their economic responsibilities only in terms of company profits and shareholder value, companies were able to ignore the problems with the new derivative markets, which, in the long-run, proved to be to their detriment.

One of the reasons why economists failed to anticipate the recent financial crisis, and why policy-makers and corporate role-players were able to define systemic risk outside of their sphere of influence and responsibility, was because of an over-reliance on '[f]ragmented, compartmentalized, mechanized, disjunctive, reductionist intelligence', which, as Morin (1999: 17) warns, 'is nearsighted and often goes blind.' By this he means that individuals, who are unable to grasp a problem in its entirety, are also unlikely to engage in reflection, indicating that 'the chances of corrective judgement or a long term view are drastically reduced.'

As is well-recognised today, the financial crisis was not only the result of personal greed (although there were a number of actors that sought to exploit the system to their own advantage), but was in the first instance made possible by the fact that the financial institution was severely weakened by unsound regulations, a poor risk management system, business interdependencies, lack of transparency, and a faulty reward system (Crane and Matten 2008) – influences which were not sufficiently accounted for by key role-players in the crisis. In the long-run, these institutions of integrity (i.e. the norms, policies, and rules) served to undermine the integrity of the financial system. Of all these factors, the regulatory environment arguably played the most significant role in facilitating the conditions for the financial crisis, and it is to businesses' responsibilities in the legal domain that we turn next.

Legal Responsibilities

With regard to the legal domain, Schwartz and Carroll (2003) argue that CSR pertains 'to the business firm's responsiveness to legal expectations mandated by society is the form of federal, state, and local jurisdictions, or through legal principles as developed in case law' (509). In this regard, legality can be viewed in terms of compliance, avoidance of civil litigation, and anticipation of law (which takes the form of voluntary adherence to probable future legislation). As previously stated, the rules of the game are fairly rigid and transparent, and companies have a duty to act within these rules. However, the legal system is also a human institution, and hence also fallible and susceptible to human error and greed. In this regard, Roy Barnes (in Morgenson and Rosner 2011), a former governor of Georgia, remarked in 2002 that '[w]hile predatory lending violates all notions of decency and ethics, it has been largely legal due to previously loose consumer protection laws. This is not only wrong – it is tragic. And it must end.' Yet, despite concerns amongst U.S. state legislators and lawyers about the increasing number of loans issued to low-income buyers, the practice continued unabated. According to Gary Burtless (2009), the '[c]lose political ties between Wall Street and the government played a sizeable role in creating a regulatory environment in which financial institutions became dangerously over-exposed to risk'. The legal and the economic domains were thus so intertwined that inevitable conflicts of interests arose.

The purpose of this discussion is not to shed light on the reasons for the financial crisis, but merely to illustrate the degree of complexity and interconnectivity that define our current day market and legal systems. Generating profits without resorting to deception, force, or fraud still remains an important business responsibility, but given the impact that business has on the regulatory and financial systems, it is also important for business to morally reflect on their role within these domains, and to accept responsibility for their role in contributing to systemic risks, even when these risks are the consequence of a number of complex factors.

Crane and Matten (2004b) concur, arguing that when one views law and ethics as essentially separate domains (and when ethics is seen as extending the legal domain, as is traditionally the case), then one negates the role that ethics and business ethics can play in shaping the law. They further state that '[t]his is particularly an issue in an increasingly globalizing business environment' (359). Instead of narrowly defining business ethics as beginning 'where the law ends' (as Linda Trevino and Katherine Nelson (2004: 14) would have it), one should account for the interdependencies between law and business, and define business ethics in terms of 'where the law begins' (Crane and Matten 2004b: 363). This same argument can be made with regard to economics: business responsibility does not only amount to compliance with the rule of the game, but should also include a critical evaluation of these rules and the systemic risks that they pose to financial markets.

Tackling the Uncertainties of CSR

As should be clear from the preceding discussion, the uncertainties and complexities that those engaging in CSR have to deal with require careful consideration at the company, industry, national, and global levels. As such, multi-level, multi-pronged responses are necessary, and our discourses and tools should reflect the complexities with which we grapple. This is generally not the case, as noted by Sargut and McGarth (2011: 70), who argue that, in terms of managing complexity, 'our analytic tools haven't kept up.'

Problems with Traditional CSR Instruments

This is unfortunately also true of popular instruments currently used in order to determine corporate responsibilities. For example, part of the success of Carroll's framework is that one can derive 'the stakeholder/responsibility matrix' (see Fig. 6.4) from the CSR pyramid. This matrix is described by Carroll (1991: 44) as 'an analytical tool or template to organize a manager's thoughts and ideas about what the firm ought to be doing in an economic, legal, ethical, and philanthropic sense with respect to its identified stakeholder groups.' Carroll's model – in which stakeholders and corporate responsibilities are framed as independent, and in which averages or medians are extrapolated to inform our understanding of entire stakeholder populations – cannot sufficiently capture the complexities that emerge from our relational and embedded contexts.

	Types of CSR			
Stakeholders	Economic	Legal	Ethical	Philanthropic
Owners				
Customers				
Employees				
Community				
Competitors				
Suppliers				
Social Activist Groups				
Public at Large				
Others				

Fig. 6.4 The stakeholder/responsibility matrix (Carroll 1991: 44)

Given a complexity-inspired perspective of CSR, it is impossible to formulate a generic template for 'calculating' corporate responsibilities, since such a tool is premised on too many assumptions that have been debunked in this study (for example, that there exists a linear relation between cause and effect, that we can objectively determine stakeholder claims, that responsibility can be fully circumscribed, and that phenomena can be studied in isolation from one another). Taking complexity seriously does not mean that we are left without any guidance or tools, but it does mean that we should develop new quantitative and qualitative tools for managing complexity.

Management 'Tools' for Complex Times

Sargut and McGarth (2011) argue that, in order to overcome the problems associated with unforeseen corporate consequences and difficulties in sense-making, it is necessary to rethink our forecasting, risk-management, and trade-off strategies and methods. Although their discussion pertains to managing corporate performance amidst the complexities that define the general business environment, their discussion nevertheless bears fruitful consequences for the management of corporate responsibilities, and in this section these general strategies and their applicability to the domain of CSR will be elaborated on.

Improved Forecasting Methods

With regard to improved forecasting methods, Sargut and McGarth offer three points of advice. Firstly, they advocate that managers drop certain forecasting tools, specifically analytic tools that frame phenomena as independent and that extrapolate averages or medians to entire populations. As previously argued, both these assumptions underlie Carroll's (1991) 'stakeholder/responsibility' matrix, in which the goal is to 'carefully and deliberately mov[e] through the various cells of the matrix', thereby helping management to 'develop a significant description and analytical data base that can then be used for purposes of stakeholder management' (44). Sargut and McGarth (2011) argue that in complex systems, what is of interest is the nature of the interconnections between systems, as well as the wide variations contained in the system. As such, our forecasting tools should allow us to reflect on interconnectivities and outliers, as these factors are likely to tell us more about both the issues and stakeholders that are likely to play a decisive role in promoting or undermining corporate sustainability.

Secondly, Sargut and McGarth suggest that management should stimulate the corporate system by using forecasting models that 'incorporate low-probability but high-impact extremes' (73). In terms of CSR, this point implies that we should take cognisance of 'the canary in the coalmine' events and heed early warning signals, since such events can have a great impact on whether corporations deal successfully with their responsibilities or not. In this regard, the BP oil spill again serves as a good example: since the probability of an oil spill was relatively low, BP did not have the necessary processes in place to deal with the disaster. As a result, BP's 'once vaunted sustainability reputation' (Crane and Matten 2010) was destroyed by this event, and the damage done to the natural environment is incalculable.

Similarly, ignoring the concerns of marginal stakeholders can also be detrimental to a company's CSR reputation. For example, it is unlikely that Nestlé would define the environmental activist group, Greenpeace, as one of its core stakeholders. Yet, in 2010, Greenpeace activists managed to do serious harm to Nestlé's reputation (Crane and Matten 2010). What made this stakeholder group's influence on the corporation even more difficult to predict was the fact that their gripe was not with Nestlé itself, but with one of its palm oil suppliers, namely Sinar Mas. Sinar Mas is an Indonesian palm oil producer, whose corporate activities have contributed to deforestation. In order to draw attention to Sinar Mas's environmentally exploitative operations, Greenpeace activists created a spoof ad for Nestlé's Kit Kat chocolate bar on YouTube, which went viral. This ad shows a tired office worker breaking open a kit-kat and tucking into one of the chocolate fingers. However, on closer inspection, one sees that it is actually an orang-utan finger that the office worker bites into! The slogan of the ad reads: 'Give the orang-utan a break...' This example demonstrates the interconnectedness of stakeholders, and gives voice to Nestlé's *inexistant* stakeholders, which in this case, are orang-utans and other forest animals.

Thirdly, Sargut and McGarth argue that companies should not only rely on predictive models developed on historical data, since implicit in these models is the belief that the future resembles the past, which is an assumption that rarely holds in complex systems. Rather, historical or 'lagging' data should be supplemented by 'current', as well as 'leading' data, where the latter is defined as 'data about where things could go and how the system might respond to a range of possibilities' (74). They further argue that leading data 'will be fuzzy and subjective by definition' (74), and here we see the value that Morin's 'logic of complexity' holds for developing leading data. In terms of CSR, one can argue that generating leading data is specifically important for developing sustainable strategies, since past and current tools are proving inadequate for dealing with the scope of our environmental, social, and economic problems.

Better Risk Mitigation

In order to deal with the 'planetary problem', and the sustainability issues that characterise this problem, we need to not only develop better forecasting strategies, but also better risk mitigation strategies. In a complex system, a measure of risk is unavoidable, simply because we cannot accurately model the system. However, Sargut and McGarth (2011) point to a number of strategies that help one to lessen the risk. Firstly, they argue that management should '[l]imit or even eliminate the need for accurate predictions' (75), because in a complex world, many things are quite simply unpredictable. Again, this reinforces the point that business responsibilities cannot be fully determined in codes or CSR policies, since it is often the unexpected that calls for action. In the first instance, CSR should be about developing an extended awareness of one's responsibilities, because only if this is the case, will one be able to develop successful strategies, which Morin (2008: 96) defines as 'the art of working with uncertainty'.

Secondly, Sargut and McGarth (2011) propose the use of storytelling and counterfactuals as a way of countering risk. Specifically, they argue that:

> Sharing anecdotes about near misses and rehearsing responses to a hypothesized negative event can help focus attention on a possibly significant future occurrence. Posing counterfactuals - asking "What if?" – is a terrific but surprisingly underutilized way of coming up with scenarios that are unlikely to be surfaced by traditional techniques (75).

In terms of CSR, storytelling and counterfactuals also present a great way of stimulating moral imagination and can help us to transgress the boundaries of our social practices, simply because 'the storyteller's reflections are not restricted by the available data' (75). Through active dialogue, these concerns can be used to create a heightened awareness of present and future risks that may arise from the organisation's activities, and which in turn focuses employee attention on the organisation's responsibilities in dealing with these potential risks.

Thirdly, Sargut and McGarth (2011) argue that the soft, flexible story-telling approach should be supplemented with rigid, quantitative analyses that are nevertheless multi-pronged. As such, triangulation – which implies the uses of a number of methodologies, assumptions and data tools – presents a valuable way of ensuring that storytelling and counterfactuals remain realistic and relevant to the specifics of the situation and for attacking complex problems from various angles. Triangulation is also a viable method for calculating risk, because more of the complexity can be taken into account due to the use of a number of methodologies and tools. In terms of CSR, this argument holds important consequences for the scope of our perceived responsibilities, since the use of a number of tools (both quantitative and qualitative) represents a means of improving on our reading strategies when engaging with the (con)text, and of reinscribing responsibility and truth in 'larger, more stratified contexts' (Derrida 1988: 146).

Smart Trade-off Decisions

No matter how well we forecast the future, or mitigate our risk, difficult decisions that involve trade-offs are inevitable. In complex environments, managing corporate responsibilities does not constitute a simple engineering problem, but requires that we grapple with contesting meanings, tensions, overdeterminations, and equivocations (Derrida 1999). In this regard, Sargut and McGarth (2011: 75–76) recommend that managers take a 'real-options approach', which means 'making relatively small investments that give you the right, but not the obligation, to make further investments later on.' This real-options approach goes hand-in-hand with Cilliers' (2006) argument for slowness. Their argument 'is against unreflective speed, speed at all cost, or, more precisely, against speed as a virtue in itself: against the alignment of "speed" with notions like efficiency, success, quality and importance' (106). Both the real-options approach and the argument for slowness emphasise the importance of careful and considered action in complex contexts. If we do not fully know what the consequences of our actions are, it is imprudent and irresponsible to act in ways that potentially have irreversible negative consequences, merely for the sake of making 'a quick buck'. Perhaps if the world moved according to a different tempo, we may have been able to limit the extent of the recent financial crisis and the current environmental crisis. In terms of CSR, corporate members not only have a responsibility to nurture an extended awareness of their responsibilities, but also to engage in a meaningful and considered manner with these responsibilities.

Lastly, Sargut and McGarth (2011) point to the importance of cognitive diversity as a means for ensuring good trade-offs. This point has been argued at length, but to recap, diversity of thought is necessary to ensure that corporations stay attuned to underlying risks and trends, and to be able to successfully deal with environmental complexities. However, successful and sustainable organisations are not only populated by diverse thinkers, but also have the necessary culture, policies, and procedures in place, in order to harness this diversity. In this regard, leadership plays an essential role in ensuring that organisations are managed in an effective and responsible manner. The analysis would therefore be incomplete without an overview of the type of leadership that supports a complex notion of CSR.

Leadership and the Development of Robust Corporate Cultures

Leadership Approaches

An overview of the extant leadership literature reveals two dominant approaches to leadership, namely the 'entity' or 'agential' approach and the 'relational' or 'systemic' approach (Uhl-Bien 2006). These approaches are described and reviewed in Grebe and Woermann (2011: 26–37), and a short summary of the argument is presented here.

The Agential Approach

The agential approach constitutes the traditional take on leadership, in which the focus is on leaders and their individual character traits, and where the goal of leadership is to influence and control followers, in order to mobilise people and resources. The agential approach can further be divided into the normative and descriptive perspectives. In the normative perspective, the focus is only on the person of the leader, whereas the scope of analysis in the descriptive perspective is extended to include the empirical determinants that give rise to successful leadership. The guiding question in the normative perspective is 'what ought successful leaders to do?', whereas those who work within the descriptive perspective, attempt to provide answers to the question 'what do successful leaders actually do?'.

Brown and Trevino (2006) are proponents of the descriptive agential approach, and convincingly argue that leadership theories must be able to account for the 'transactional' (600) or behavioural aspects of leadership. What is specifically important in this regard, is the way in which leaders can help to create, reinforce, or change an institution's culture or direction (Schein 1985). However, even within the descriptive perspective, the leader is still understood as 'the architect and controller of an internal and external order' (Uhl-Bien 2006: 65). As such, the agential approach to leadership focuses on agency at the expense of structural and systematic considerations. The systemic leadership approach, on the other hand, is capable of accounting for both structure and agency, and is therefore compatible with the embedded, relational, and emergent nature of corporate identity and responsibility developed in this chapter.

The Systemic Approach

The systemic approach, which is currently gaining prominence in the business ethics and leadership literature, views both leaders and leadership processes as emergent properties of a system (see Hosking 2007; Uhl-Bien 2006; Collier and Esteban 2000). Unlike the agential approach, the systemic approach is also sensitive to the contextual contingencies and dynamic stakeholder relations that impact on leaders and leadership processes, as is clear from the example of Barak Obama's rise to presidency. Instead of analysing Barak Obama's rise to the presidency in terms of his individual character traits (the normative agential approach), or in terms of behavioural attributes or choices undertaken by a man that is in control of his destiny (the descriptive agential approach), what is relevant in terms of the systemic approach to leadership, are the complex, social processes (including the US culture, the Democratic Party machinery, the intense desire for change, an overwhelming dislike of the Bush administration, the importance of internet campaigning, the credit crisis etc.) and the affect that these processes had on the voting public (specifically key stakeholder such as black Americans, environmentalists, mainstream democrats, a younger generation of well-informed techno-geeks, and disenchanted Republicans) (Grebe and Woermann 2011).

Grebe and Woermann (2011) argue that, although this example is useful in elucidating some of the characteristics of the systemic approach, it fails to capture the distributed nature of systemic leadership. In this approach, leadership is often 'stretched over the practice of actors within organizations' (Friedman 2004: 206) and 'many of the functions that were traditionally associated exclusively with formal leadership are now shared by members of an organizational system' (Painter-Morland 2008: 229). What this means is that the process of mobilising people and resources can only be understood in terms of an institution's systemic functions. This is not to say that individuals cannot assume leadership roles. The point is rather that different individuals can embody leadership roles at different times, depending on the circumstances (Painter-Morland 2008).

Following Peter Gronn, Grebe and Woermann (2011) further argue that distributed leadership can be understood in two ways: firstly, the common, numerical understanding of distributed leadership views the leadership function as dispersed amongst some, many, or all members of an organisation (Wenger 2000). On this reading, distributed leadership is the sum of the attributed influence of organisational members. Uhl-Bien et al.'s (2007) discussion on the differences between administrative, adaptive, and enabling leadership serves as a good example in this regard. Secondly, distributed leadership can be understood as joint or concertive action (Gronn 2003). Distributed leadership, as concertive action, implies that:

> There are collaborative models of engagement that arise spontaneously in the workplace.
>
> There is the intuitive understanding that develops as part of close working relations between colleagues.
>
> There are a number of structural relations and institutionalised arrangements which constitute attempts to regularise distributed actions (35).

This latter understanding of distributed leadership accords well with a complexity-view of the organisation, since it not only accounts for the fact that the leadership function is stretched across individuals in the organisation, but also for the fact that organisational nodes or coalitions (that can play an influential role in determining the direction of the organisation) emerge from cooperative and competitive organisational activities.

The above offers only a very short introduction to systemic leadership. What is more important for our purposes are the implications that systemic leadership holds for nurturing and promoting responsible action, and in this regard it is argued that leaders should undertake a number of functions that can help to promote organisational openness and responsiveness – both of which are critical for safeguarding responsible corporate action. The discussion on these leadership responsibilities again draws from the analysis given in Grebe and Woermann (2011).

Leadership Responsibilities

Leaders Should Guide the Process of 'Organisational Becoming'

Since the systemic approach to leadership is premised on the idea that our identities (and, hence, our sense of right and wrong) emerge within communities of practice, an important leadership responsibility is to guide the organisational 'process of becoming' (i.e. the process by which an institution's or collective's identity evolves, or is constituted, through the establishment of shared meaning). In the previous chapter, a number of examples of differences between cultural contexts were given, as part of an argument against a universal or *given* ethics. What these examples showed was that what 'ought' morally-speaking to be the case (the normative dimension) is, in part, determined by considerations of what is the case (the descriptive dimension). In other words, our sense of normative propriety is influenced by our contexts. However, what Derrida's philosophy teaches us is that our contextually-formed practices must be continually subjected to deconstructive analyses; since it is only by challenging the status quo that we are able to dismantle certain undesirable institutional practices. Responsible thought and action is therefore inextricably linked with the desire to do justice, and to remain open to the future.

In the corporate environment, leaders have the important task of promoting organisational openness, so that organisations remain aware of their responsibilities and remain responsive to their environments. The reason why this task falls largely to leaders, is because both formal and informal leaders occupy positions of power, and are therefore capable of influencing the institutions of integrity. In this regard, Etienne Wenger (1998: 191) argues that systemic leadership 'nurtures community', and, that in this role, systemic leadership 'is deeply ethical because it is the process of belonging to a community which is constitutive of identity.' However, despite the critical role that leaders play in nurturing community, it must also be noted that, contrary to what is assumed in the agential approach, leaders cannot exclusively determine or control institutional identity. This is because no single actor is capable of mastering or controlling contextual complexities, and this fact also explains why leaders have the responsibility to attend to broader institutional narratives.

Leaders Should Attend to Emergent Institutional Narratives and Sensitise Institutional Role-Players to the Effects of Their Actions

The distributed nature of systemic leadership implies that 'leadership qualities of competence, judgement and decision-taking are needed throughout the [organisation]' (Collier and Esteban 2000: 207). Leadership is thus 'an on-going direction-finding process, which is innovative and continually emergent' (208), and, in this regard, individual leaders can play an important sense-making role. Leadership strategies that are moulded on 'those aspects that 'fit' some preconceived picture' (Lissack and Letiche 2002: 91) are doomed to fail, because institutional narratives constitute the shared meanings or 'stories about ourselves' that emerge from a continuous and complex process of dialogue and interaction. Sense-making requires that leaders engage in this dialogue and take cognisance of emergent organisational themes so that they can guide, intervene in, and attend to, organisational narratives, in order to ensure that the shared meanings arrived at are appropriate and inclusive (i.e. congruent with institutional rules and norms, and with environmental demands).

Leaders also have the related responsibility of sensitising organisational members to the effects that their actions have on stakeholders. Leaders can achieve this by drawing attention to the network of relations to which an organisation belongs, and which is critical to its survival and identity. Stakeholders form an important part of this network, and should not be seen as a nuisance to be managed, but as integral to the well-being and flourishing of the organisation itself. However, it is only possible to sufficiently account for the well-being of stakeholders if we are aware and capable of making sense of our larger, emergent organisational identities, since this extended awareness will allow us to better understand the interests and concerns of stakeholders.

Leaders Should Foster Institutional Congruence

Through showing a preparedness to attend to institutional narratives, to engage in dialogue, and to remain sensitive to stakeholder needs – and, through encouraging other institutional role-players to do the same – leaders can help to foster institutional congruence. Earlier in this chapter, congruence was defined as functional unity or sense of purpose. Congruence is the process of aligning internal and external stakeholder expectations and actions with institutional or cultural norms and codes, in order to foster institutional purpose. It is important to note that congruence is not a superficial or retrospective experience of institutional 'sameness'. Rather, it is the outcome of experimentation and contention, and allows systems to draw on the full range of talents, skills, and perspective that diverse stakeholders have to offer (Painter-Morland 2008). Although preferring the term 'coherence' over that of 'congruence', Lissack and Letiche (2002) provide further insight into what congruence might mean: it implies a shared meaning and signifying apparatus; it demands robustness; it is about forging a powerful link between people and circumstances; it is enabling; and, it includes and confronts emergence.

Congruence is therefore that which prevents the outcomes of cognitive diversity from degenerating into noise – because, despite the range of talents, opinions, interests, claims, and goals that may exist within an organisational network, the system itself must retain its functional integrity. Without functional integrity organisations will not be able to meet even their most basic responsibilities pertaining to organisational sustainability.

Leaders Should Manage Institutional Conflict and Promote Organisational Openness

In order to foster congruence, leaders should not only attend to emergent institutional narratives, but should also promote the constructive handling of institutional conflict. Since congruence is not the same as consensus, disagreements and conflicts will arise. However, such conflicts should not be allowed to degenerate into something negative, as this will undermine the integrity of institutions. In order to successfully deal with conflict, leaders should encourage dialogue and participation, as well as exercise (and encourage role-players to exercise) self-reflection. Self-reflection means being prepared to challenge one's own convictions and presuppositions (Collier and Esteban 2000), and here again we see the important role that deconstruction can play in helping us to reflect on our own biases, as well as on staid patterns of organisational authority and hierarchy. However, such a critical reflection is only likely to have positive outcomes if openness and tolerance characterise the organisation's culture. Practicing tolerance goes hand-in-hand with the acceptance of human diversity, since, if we do not respect this rich diversity, we are unlikely to respect 'the rights of others to express different or even opposite choices, convictions, and ideas' (Morin 1999: 54). In terms of CSR, reflection also helps to guard against 'universal' notions of ethics and corporate responsibility, by allowing us to see more of the complexity than would otherwise be the case.

However, whilst leaders should encourage creativity and new ideas, these ideas should be managed in a disciplined manner, because not all ideas can be implemented (Collier and Esteban 2000). Therefore, leaders should try to ensure that new ideas are broadly aligned with institutional purpose. If this is not the case, the actions of individual role-players can serve to undermine the integrity of institutions, and hence the institutions of integrity. Diversity, in other words, should be to the advantage, and not at the expense of, the organisation.

Future Challenges

The theory and model of CSR developed in this chapter not only hold important consequences for management and leadership practices, but also for a number of business ethics themes related to corporate social responsibility, such as corporate accountability (which is mostly still understood in causal terms, where corporate members or corporations are held accountable for failing to meet a number of specified responsibilities), corporate wrongdoing (in which individual moral scoundrels are sought out for blame), and corporate governance (in which business responsibilities are often presented in terms of a number of specific rules or principles).

As with traditional conceptions of CSR, these themes are based on a specific view of moral agency, and do not sufficiently account for the impact that contextual factors have on our practices. By drawing on the ideas of deconstruction and complexity theory, and by appealing to examples from the world of business, it was demonstrated in this chapter that our commonly-held notions of freedom, intentionality, and causality do not correspond with our lived experiences, and that the relational, emergent, and embedded nature of our contexts (business and otherwise) is central to determining our understanding of our responsibilities.

As such, the reconfiguration of CSR offered here is illustrative of a larger challenge that complexity thinking holds for the paradigm of business ethics, since, as Morin (2008: 34) argues, '[w]hat affects a paradigm, that is, the vault key of a whole system of thought, affects the ontology, the methodology, the epistemology, the logic, and by consequence, the practices, the society, and the politics.'

It is beyond the scope of this study to explore the above-mentioned business ethics themes. However, a last issue to which we turn is the implications that this study holds for teaching business ethics. If business ethics is to play a larger role in our business practices, then it is essential that future business practitioners and leaders develop an understanding of the importance of business ethics. As argued by Crane and Matten (2004b: 366): the challenge to moving past the current business ethics curriculum 'is a major one, it is also a timely and exciting one that may ultimately serve to maintain the relevance and revelatory potential of the business ethics subject into the future.'

References

Albert, S., B.E. Ashforth, and J.E. Dutton. 2000. Organizational identity and identification: Chartering new waters and building new bridges. *Academy of Management Review* 25(1): 13–17.

Allen, P. 2001. A complex systems approach to learning, adaptive networks. *International Journal of Innovation Management* 5(2): 149–180.

Badiou, A. 2009. *Pocket pantheon*, trans. D. Macey. London: Verso.

Brown, M.E., and L.K. Trevino. 2006. Ethical leadership: A review and future directions. *The Leadership Quarterly* 17: 595–616.

Burtless, G. 2009. The financial crisis and the rule of law. *The Brookings Institute*. Available online at: http://www.brookings.edu/opinions/2009/0414_financial_crisis_burtless.aspx

Carroll, A.B. 1991. The pyramid of corporate social responsibility: Toward the moral management of organizational stakeholders. *Business Horizons*, July-August 1991: 39–48.

Cilliers, P. 2001. Boundaries, hierarchies & networks in complex systems. *International Journal of Innovation Management* 5(2): 135–147.

Cilliers, P. 2006. On the importance of a certain slowness. *Emergence: Complexity and Organization* 8(3): 105–112.

Cilliers, P. 2008. Complexity theory as a general framework for sustainability science. In *Exploring sustainability science: A Southern African perspective*, ed. M. Burns and A. Weaver, 38–57. Stellenbosch: African Sun Media.

Cilliers, P. 2010. Difference, identity, and complexity. *Philosophy Today*, Spring 2010: 55–65.

Colander, D., M. Goldberg, A. Haas, K. Juselius, A. Kirman, T. Lux, and B. Sloth. 2009. The financial crisis and the systemic failure of the economic profession. *Critical Review* 21(2–3): 249–267.

Collier, J., and R. Esteban. 1999. Governance in the participative organisation: Freedom, creativity and ethics. *Journal of Business Ethics* 21: 173–188.

Collier, J., and R. Esteban. 2000. Systemic leadership: Ethical and effective. *The Leadership and Organization Development Journal* 21(4): 207–215.

Crane, A., and D. Matten. 2004a. *Business ethics: A European perspective. Managing corporate citizenship and sustainability in the age of globalisation*. Oxford: Oxford University Press.

Crane, A., and D. Matten. 2004b. Questioning the domain of the business ethics curriculum. *Journal of Business Ethics* 54: 357–369.

Crane, A., and Matten, D. 2008. Ethics and the financial crisis. *Crane and Matten Blog*, Thursday, 18 Sept 2008. Available online at: http://craneandmatten.blogspot.com/2008/09/ethics-and-financial-crisis.html

Crane, A. and Matten, D. 2010. Top 10 corporate responsibility stories of 2010. *Crane and Matten Blog*, Monday, 13 Dec 2010. Available online at: http://craneandmatten.blogspot.com/2010/12/top-10-corporate-responsibility-stories.html

Derrida, J. 1988. Afterword. In *Limited Inc*, ed. G. Graff and trans. S. Weber, 111–160. Evanston: Northern Western University Press.

Derrida, J. 1995. "Eating well," or the calculation of the subject. In *Points... Interviews, 1974–1994*, ed. E. Weber and trans. P. Kamuf, 255–287. Stanford: Stanford University Press.
Derrida, J. 1997. The Villanova roundtable. A conversation with Jacques Derrida. In *Deconstruction in a nutshell: A conversation with Jacques Derrida*, ed. J.D. Caputo, 1–30. New York: Fordham University Press.
Derrida, J. 1999. Hospitality, justice and responsibility: A dialogue with Jacques Derrida. In *Questioning ethics: Contemporary debates in philosophy*, ed. R. Kearney and M. Dooley, 65–83. London/New York: Routledge.
Derrida, J. 2002. Negotiations. In *Negotiations: Interventions and interviews, 1971–2001*, ed. and trans. E. Rottenberg, 11–40. Stanford: Stanford University Press.
Donaldson, T., and L.E. Preston. 1995. The stakeholder theory of the corporation: Concepts, evidence, and implications. *Academy of Management Review* 20(1): 65–91.
Erskine, T. 2003. *Can institutions have responsibility? Collective moral agency and international relations*. London: Palgrave.
Floridi, L. 1999. Information ethics: On the philosophical foundation of computer ethics. *Ethics and Information Technology* 1(1): 37–56.
Freeman, R.E. 2008. Managing for stakeholders. In *Ethical issues in business: A philosophical approach*, ed. T. Donaldson and P.H. Werhane, 39–53. Upper Saddle River: Pearson.
Freeman, R.E., and R.A. Philips. 2002. Stakeholder theory: A libertarian defense. *Business Ethics Quarterly* 12: 331–349.
French, P.A. 1979. *Collective and corporate responsibility*. New York: Columbia University Press.
French, P.A. 1984. The corporation as a moral person. *American Philosophical Quarterly* 16: 207–215.
Friedman, M. 1962. *Capitalism and freedom*. Chicago: University of Chicago Press.
Friedman, A.A. 2004. Beyond mediocrity: Transformational leadership within a transactional framework. *International Journal of Leadership in Education* 7(3): 203–224.
Gioia, D.A., M. Schultz, and K.G. Corley. 2000. Organizational identity, image, and adaptive instability. *Academy of Management Review* 25(1): 63–81.
Grebe, E. and Woermann, M. 2011. Institutions of integrity and the integrity of institutions: Integrity and ethics in the politics of developmental leadership. Research Paper. *Developmental Leadership Program*. March 2011: 26–37.
Gronn, P. 2003. *The new work of educational leaders: Changing leadership practices in an era of school reform*. London: Paul Chapman Publishing.
Hattingh, J.P. 2006. The state of the art in environmental ethics as a practical enterprise: A view from the Johannesburg documents. In *Environmental ethics and international policy*, ed. H.A.M.J. ten Have, 191–216. Paris: UNESCO Publication.
Homann, K. 2007. Globalisation from a business ethics point of view. In *Globalisation and business ethics*, ed. K. Homan, P. Koslowski, and C. Luetge, 3–10. Hampshire/Burlington: Ashgate Publishing.
Hosking, D.M. 2007. Not leaders, not followers: A post-modern discourse of leadership processes. In *Follower-centered perspectives on leadership: A tribute to the memory of James R. Meindl*, ed. B. Shamir, R. Pillai, M. Blighand, and M. Uhl-Bien, 243–264. Greenwich: Information Age Publishing.
Keeley, M. 1981. Organizations as non-persons. *Journal of Value Inquiry* 15: 149–155.
Keeley, M. 1988. *A social-contract theory of organizations*. Notre Dame: University of Notre Dame.
Ladd, J. 1970. Morality and the ideal of rationality in formal organizations. *The Monist* 54(4): 488–516.
Lissack, M.R., and H. Letiche. 2002. Complexity, emergence, resilience, and coherence: Gaining perspectives on organizations and their study. *Emergence* 4(3): 72–94.
Mauboussin, M.J. 2011. Embracing complexity. *Harvard Business Review*, September 2011: 89–92.
Midgley, G. 2003. Science as systemic intervention: Some implications of systems thinking and complexity for the philosophy of science. *Systemic Practice and Action Research* 16(2): 77–97.

Mitchell, R.K., B.R. Agle, and D.J. Wood. 1997. Towards a theory of stakeholder identification and salience: Defending the principle of who and what really counts. *Academy of Management Review* 22(4): 853–886.

Morgenson, G., and J. Rosner. 2011. *Reckless endangerment: How outsized ambition, greed, and corruption led to economic Armageddon*. New York: Times Books.

Morin, E. 1999. *Seven complex lessons in education for the future*, trans. N. Poller. Paris: UNESCO Publication.

Morin, E. 2007. Restricted complexity, general complexity, trans. C. Gershenson. In *Worldviews, science and us: Philosophy and complexity*, ed. C. Gershenson, D. Aerts, and B. Edmonds, 5–29. Singapore: World Scientific.

Morin, E. 2008. *On complexity*, trans. S.M. Kelly. Cresskill: Hampton Press.

Painter-Morland, M. 2006. Redefining accountability as relational responsiveness. *Journal of Business Ethics* 66: 89–98.

Painter-Morland, M. 2008. *Business ethics as practice: Ethics as the everyday business of business*. Cambridge: Cambridge University Press.

Petit, P. 2007. Responsibility incorporated. *Ethics* 117: 171–201.

Polonsky, M., and C. Jevons. 2006. Understanding issue complexity when building a socially responsible brand. *European Business Review* 18(5): 340–349.

Richardson, K.A., and M.R. Lissack. 2001. On the status of boundaries, both national and organizational: A complex systems perspective. *Emergence* 3(4): 32–49.

Salancik, G.R., and J. Pfeffer. 1974. The base and use of power in organizational decision-making: The case of universities. *Administrative Science Quarterly* 19: 453–473.

Sargut, G., and R.G. McGarth. 2011. Learning to live with complexity. *Harvard Business Review*, September 2011: 69–76.

Schein, E. 1985. *Organizational culture and leadership*. San Francisco: Jossey-Bass.

Schwartz, M.S., and A.B.. Carroll. 2003. Corporate social responsibility: A three-domain approach. *Business Ethics Quarterly* 13(4): 503–530.

Sims, R.R., and J. Brinkmann. 2003. Enron ethics (or: Culture matters more than codes). *Journal of Business Ethics* 45: 243–256.

Siponen, M. 2004. A pragmatic evaluation of the theory of information ethics. *Ethics and Information Technology* 6(4): 279–290.

Smith, A. 1776. *An inquiry into the nature and cause of the wealth of nations*. London: W. Strahan and T. Cadel.

Trevino, L.K., and M.E. Brown. 2004. Managing to be ethical: Debunking five business ethics myths. *The Academy of Management Executive* 18(2): 69–81.

Trevino, L.K., and K.A. Nelson. 2004. *Managing business ethics: Straight talk about how to do it right*, 3rd ed. New York: Wiley.

Uhl-Bien, M. 2006. Relational leadership theory: Exploring the social processes of leadership and organizing. *The Leadership Quarterly* 17: 654–676.

Uhl-Bien, M., R. Marion, and B. McKelvey. 2007. Complexity leadership theory: Shifting leadership from the industrial age to the knowledge era. *The Leadership Quarterly* 18: 298–318.

Velasquez, M.G. 1983. Why corporations are not morally responsible for anything they do. *Business and Professional Ethics Journal* 2(3): 1–18.

Wenger, E. 1998. *Communities of practice: Learning, meaning and identity*. Cambridge: Cambridge University Press.

Wenger, E. 2000. Communities of practice and social learning. *Organizations* 7(2): 225–246.

Werhane, P.H. 1980. Formal organizations, economic freedom and moral agency. *Journal of Value Inquiry* 14: 43–50.

Woermann, M. 2010. Corporate identity, responsibility and the ethics of complexity. In *Complexity, difference and identity*, ed. P. Cilliers and R. Preiser, 167–192. Dordrecht: Springer.

Woermann, M. 2011. In corporations we trust? A critique of contractarian-based corporate social responsibility models. *African Journal of Business Ethics* 5(1): 26–37.

Chapter 7
Implications for Teaching Business Ethics

Abstract In this concluding chapter, an overview is given of the core components that should be addressed in business ethics education, in order to both deal with the fundamental problems that characterise our times, as well as to promote the future viability of business ethics. The analysis of these core components is based on the complexity and Derridean insights presented in the foregoing chapters, and the aim of the analysis is to unpack teaching strategies that can equip students with the sense-making tools and tools of analysis needed to reflect upon the normative dimensions of complex business challenges. Since these challenges are context-dependent, the analysis does not provide examples of specific pedagogical interventions, because these interventions must be forged within specific environments. The suggestions made in this chapter therefore stand prior to any particular teaching guide or curriculum.

Introduction

In the first chapter of this study, it was argued that the goal of business ethics should be to provide students and practitioners with sense-making tools and tools of analysis that can aid in ethical decision-making in the workplace, and the first part of the study concluded with the insight that our analytic tools should be supplemented with other tools to help us to better account for, and deal with, the complexities that define our contexts. A number of alternative tools have already been introduced in this study; but, at this juncture, we turn specifically to knowledge, which can be defined as a powerful theoretical tool, in order to further investigate the type of considerations that should inform a business ethics education.

Before proceeding with the analysis, it is worthwhile to take note of Lissack and Letiche's (2002) analogy between physical tools and knowledge. Firstly, both tools and knowledge can only be understood through use, and using them modifies the user's perspective on the world. Secondly, learning how to use tools and knowledge involves a great deal more than can be captured in explicit rules. And,

M. Woermann, *On the (Im)Possibility of Business Ethics*, Issues in Business Ethics 37,
DOI 10.1007/978-94-007-5131-6_7, © Springer Science+Business Media Dordrecht 2013

thirdly, the occasions and conditions for use emerge from, and are framed within, a specific context. What this analogy implies is that, in order for our knowledge and ideas to be useful, they need to be put to use. As previously argued, from a complexity viewpoint, there is no radical distinction between ivory tower ethics and boardroom ethics, which means that our theories have the power to transform our practices, but our practices can also modify and shape our ideas.

Many business ethicists feel that complex conceptions of ethics (which are critical, provisional, and transgressive in nature) won't find footing in the 'real world', since such theories cannot provide ready solutions to the problems experienced by businesspeople. However, using Lissack and Letiche's analogy, one can argue that just as tools assist us in solving our problems, so too do theories. Despite being crucial to the task at hand, a paintbrush cannot give us a painting anymore than a violin can give us music. Rather, these tools are only of value if used by someone who possesses the requisite skills, knowledge, and imagination. The same applies to our theories. What is of crucial importance in this regard is that we use our theories in order to become competent and critical readers of texts and contexts.

It cannot be denied that complexity thinking demands more from the user than do ethical theories that provide 'moral recipes' for attaining the good life. However, in this study it was shown that notions of ethics that are justified either in terms of transcendental ideals or the functionalist paradigm are often not practically very useful, since they are premised on moral free will and on an objective, stable view of the world, respectively. Neither of these premises holds, given the complexities of the world. Successfully engaging with moral problems requires skills such as critical self-reflection, moral imagination, an ability to view the interdependent relations between parts and whole, an ability to understand subjects within their contexts, an ability to engage in dialogue and practice openness, and an ability to think through complex challenges. These are difficult skills to master and presuppose a high level of moral development, which runs contrary to the commonly-held view that being ethical is easy, and that the ethicist's work is to motivate people to make ethical choices, which will necessarily result in good consequences. As has been demonstrated in this study, this latter view unfortunately does not represent a realistic take on the nature of ethical actions. As Bauman (1993: 15) says, conceding to the complexities is unlikely to make life any easier. The best we can hope for is that by putting these ideas to work, we may make life 'a bit more *moral*.'

In this concluding chapter, ways in which to develop the above-mentioned skills will be explored in more detail. The principal text that will inform this discussion on teaching business ethics is Morin's (1999) report for the United Nations Educational, Scientific, and Cutural Organization (UNESCO), entitled 'Seven lessons in education for the future'. In the forward to this document, Morin (1) writes that his text 'stands prior to any suggested guide or curriculum' and that '[t]he intention is simply to identify fundamental problems that are overlooked or neglected in education, and [that] should be taught in the future'. One of the central insights from complexity thinking concerns the important role that context and contingency play in shaping and informing our ethical sensibilities. This means that a complex view of ethics can never prescribe to us how we should act, as each situation calls for fresh judgement. Therefore, although these fundamental problems of which Morin speaks will be contextualised in terms of the skills that should be nurtured in business ethics education, this analysis will also not provide examples of specific pedagogical interventions, since these are largely context-dependent issues. Deconstruction and complexity thinking do not conclusively provide us with 'The Answer' when attempting to specify our business responsibilities, since the concept of the good is neither a self-sustaining or natural given. Therefore, instead of presenting a prescriptive or substantive framework of what the business ethics curriculum should entail, the aim of this analysis is rather to focus awareness on the components that should be considered in any curriculum in which the complexities of business are addressed.

The Status of Our Theoretical Tools

A critical question that arises in this regard is whether deconstruction and complexity thinking are the *only* tools that can help us to successfully address the fundamental problems that characterise our times. Staying with Lissack and Letiche's analogy between tools and knowledge, one can argue that beautiful paintings, for example, need not be created with paintbrushes – sponges, or even fingers can serve as adequate replacements. The same holds true in the case of our theories: alternative theories may prove just as useful for achieving our ends. In order to be consistent with the position developed in this study, one has to concede to the possibility that other theories may be just as appropriate, or even more appropriate, in making an argument for a complex ethics. Therefore, not only do deconstruction and complexity thinking not provide us with 'The Answer', but they also cannot exclusively determine 'The Way'.

In this regard, one could claim that the choice of theories presented here is quite arbitrary. However, such an argument stems from a position akin to that of skeptical postmodernism. A better alternative is to follow a productive, affirmative, and more robust line of reasoning, and instead argue that – despite conceding to the applicability of utilising alternative theories – what is of greater importance is whether our chosen theories prove useful, and whether we employ these theories with integrity (that is, whether we show a willingness to engage with our ideas, and submit them to tests of truth and error). It is therefore of central importance that we display an attitude of critical self-reflexivity at all times, in order to prevent our ideas from controlling us (Morin 1999).

We will only be able to successfully challenge our ideas if we understand these ideas, as well as the potential power that they hold for our practices. A certain level of competence is therefore needed on the part of ethicists, students, and business practitioners. To see why this is the case, one need only consider the damage done by the jargon-filled and confusing positions espoused by certain avowed 'deconstructionists'. This point can also be made by once more turning to the analogy between tools and knowledge: although the sounds of the violin (for example) can be beautiful, in the wrong hands the violin emits a terrible, screeching noise. Similarly, our theories and ideas need to be appropriate to the relevant contexts in which they are employed. Just like an amateur musician is likely to create a more pleasant sound by strumming a few, simple guitar chords than by attempting a sophisticated piece on the violin; so too, one may achieve greater success with business ethics students without making explicit reference to deconstruction and complexity thinking – particularly because many of these students do not have any formal training in philosophy.

A fallacy which is often made is that one can *only* speak of complex phenomena in a complex language. Although one must be mindful of the simplifying effects that can be created by language, it is possible to convey the insights that emerge from this study in other ways. A specific insight highlighted in the paradigm of critical complexity that one would, for example, do well to convey in a teaching context concerns the relational and emergent nature of identity and of our systems of meaning. Furthermore, important insights pertaining to deconstruction that (I argue) should also inform our teaching strategies, include the function of the double-movement (which necessitates that we carefully interpret, but also destabilise our conceptual schema), the impetus of deconstruction (which is characterised by ethical testimony and the desire to do justice), and the logic of deconstruction (which functions as a recursive modality).

Although we might have to modify our theoretical tools in a given context, in order to make them appropriate for a given audience, the type of considerations highlighted above are nevertheless imperative for understanding the ethical position developed in this study; because, together, these various elements reinforce the contingent nature of our theoretical paradigms and, thereby, highlight the need to continually and critically engage in the ethics of complexity. Successfully translating and conveying these considerations into an appropriate 'language' for students and practitioners, whilst avoiding a dumbing down of these ideas presents an enormous challenge for ethicists. Otherwise stated, this challenge amounts to tailoring our teaching strategies to specific circumstances, whilst maintaining the integrity of the ideas with which we are working.

Components of a Business Ethics Education

In their article entitled 'Yes, you can teach business ethics: a review and research agenda', Scott Williams and Todd Dewett (2005) argue against the pessimistic view that teaching business ethics is not a worthwhile endeavour. They specifically seek to debunk three arguments against business ethics teaching. The first argument is that business ethics cannot be taught since values have already been formed prior to starting with tertiary education. Citing a number of research studies, Williams and Dewett conclude that individuals continue to develop morally throughout adolescence and young adulthood. From a complexity viewpoint, moral development – as with identity – is a fluid and emergent concept, and is characterised in terms of a state of becoming, rather than of attainment. The second argument against business ethics education is based on the separation thesis: 'Business is successful because it is driven by self-interest; ethics is simply not important or relevant in a business context' (110). Following Friedman (1962) and Smith (1776), Williams and Dewett (2005) argue that self-interested behaviour should not be equated with a disregard for others and with selfishness, since corporate performance and sustainability is contingent on operating within the rules of the game, including rules pertaining to elementary morality. They conclude in stating that 'business ethics and competitive, market-based economics are compatible in many ways' (111). Again, one can also make an argument against this criticism from a complexity-based perspective: as was demonstrated in this study, every decision is characterised by both a normative and a descriptive dimension, and in order to function both ethically and effectively, this normative dimension should be accounted for. Thirdly, Williams and Dewett cite the criticism pertaining to the relative failure of ethics training in curbing corporate scandals. As mentioned in the first chapter, the institutionalisation of courses in business ethics was, in many cases, a knee-jerk reaction to the business and accounting scandals that surfaced in the past three decades. Williams and Dewett rightly note that this critique stems from misplaced expectations regarding the goal of business ethics, since such critics fail to acknowledge the complex factors that impact on ethical behaviour. In this regard, they state that:

> Ethical behavior is a function of a variety of personal and situational factors including moral development, norms, coercion, regulations, self-control, and ethics training (Trevino 1986). No single part of the complex system of interactions among these components can control behavior by itself; each component serves an important role in creating and developing the system and helping to create circumstances that promote ethical behavior. Business ethics education must be understood in that context (112).

This description of ethical behaviour is compatible with the position developed in this study, and on the basis of this description Williams and Dewett argue that business ethics education should seek to 'increas[e] students' awareness of the ethical implications of their actions, promot[e] students' moral development, and promot[e] students' ability to handle the complexity of ethical situations' (112). These three goals constitute a useful framework for further exploring Morin's ideas, and will thus be employed to structure the analysis.

Enhancing Moral Awareness

Nurturing students' moral awareness of, and sensitivity towards, the ethical consequences of their actions is, as Williams and Dewett (2005) note, one of the most common goals of business ethics education. However, in a complex world, this is not an easy goal to attain, because as Trevino and Brown (2004: 70) state:

'Rarely do decisions come with waving red flags that say, "Hey, I am an ethical issue. Think about me in moral terms!"' Furthermore, a research review on cognitive complexity (Streufert and Nogami 1989) referenced by David Swenson and Rigoni (1999: 577) indicates that less cognitively complex persons tend 'to engage in search behavior less when there was an information overload', which is always the case in complex situations. When individuals are not aware of the normative dimensions of the issue at hand, 'moral judgment processes are not initiated' (Trevino and Brown 2004: 70), which means that individuals do not account for the consequences of their actions.

A willingness to consider ethical problems from complex perspectives is, in part, influenced by learning (although cognitive styles also play a role). In order to enhance moral awareness, and to facilitate a broader understanding of our responsibilities towards one another, business ethicists should focus on teaching what Morin (1999: 15) calls 'pertinent knowledge'. Pertinent knowledge confronts complexity, especially the type of complexity that arises when various elements (e.g. economic, political, sociological, psychological, ethical etc.) that compose a whole are inseparable. Pertinent knowledge is therefore knowledge of the complex 'inter-retroactive, interactive, interdependent tissue between the parts and the whole, the whole and the parts, [and] the parts amongst themselves' (15). In other words, pertinent knowledge concerns an understanding of the multi-dimensionality of human experiences, in which we do not isolate the parts from the whole, or isolate the parts from each other in our thinking. As previously stated, Morin (17) also argues that isolated thinking leads to a'[f]ragmented, compartmentalized, mechanized, disjunctive, reductionist intelligence', which 'is nearsighted and often goes blind', thereby reducing the chances of corrective judgement. We can only take cognisance of the wager involved in a decision, and implement (and when necessary, modify) a strategy, when we confront the complexities of our systems. Complex thought allows us to surmount short or medium term uncertainties, whilst remaining vigilant of the fact that we cannot claim to have eliminated uncertainty in the long term.

In a business ethics context, pertinent knowledge implies a focus on the coterminous formation of individual and corporate identities, corporate identities and socio-cultural identities, and the ethical implications that arise from a broadened understanding of our work practices. Acknowledging the complexity of the human condition also leads to a denouncement of the arrogant and egocentric implications that arise from subscribing to the Cartesian dogma, wherein humans are seen as isolated actors and decision-makers. If a discipline such as business ethics operates as a self-enclosed realm (taught in isolation from other subjects), then the mind begins to lose its natural aptitude 'to contextualize knowledge and integrate it into its natural entities' (16). In Montuori's (1996: 58) words: 'The segmented, fragmented organization of education into small compartments, each engaged in the study of a rigidly defined discipline, leads to blinkered assembly-line overspecialization.' Applied to the business context, this means that business activities will be viewed as standing apart from the totality of systems to which they in fact belong. Morin (1999: 16) notes that the fragmentation of knowledge has detrimental consequences, because '[a] weakened perception of the global leads to a weakened sense of responsibility (each individual tends to be responsible solely for his specialized task) and weakened solidarity (each individual loses the feeling of his ties to fellow citizens).'

In order to promote pertinent knowledge and guard against the fragmentation in education, pedagogues should focus on teaching strategies that enhance awareness of the 'mutual relations and reciprocal influences between parts and the whole in a complex world' (2). This, in turn will also help students to 'grasp subjects within their contexts, their complex, their totality' (1). Business ethics should be advanced in its own right as part of a comprehensive business curriculum (Swenson and Rigoni, 1999: 10); but, in order to achieve greater integration of ethical concerns, attention should also be given to the ethical dimensions of business and economics studies, and business ethics should be contextualised in terms of wider business realities. As such, it is necessary to cultivate a stronger interdisciplinary focus in business ethics, since how we conceptualise our practices affects the perceived scope of our business responsibilities.

Raising awareness of the normative dimension of all business practices also helps to dismantle determinist and reductionist views of ethics, since integrating ethics into our practices necessitates a more complex understanding of ethics. In this regard, deconstruction can serve as a valuable tool for helping us to engage with our embedded, differentially-defined responsibilities, because deconstruction both necessitates that we take seriously our shared and dominant traditions, and that we critically reflect on these traditions. As such, business ethicists would do well to try and nurture a deconstructive mindset in the classroom.

An important outcome of pertinent knowledge is that it enhances cognitive complexity, which helps individuals to 'tolerate apparent inconsistencies and conflicts in information, avoid extreme judgments, more accurately predict outcomes, avoid dualistic categorizing, search for more diverse information, and entertain more questions about an event' (Swenson and Rigoni 1999: 577). Even though the implications arising from the complex nature of our myriad interactions and associations with one another can never fully be grasped, our responsibility, as teachers and business practitioners, resides in engaging with these complexities and the incalculable possibilities and consequences that they may hold for our practices.

Enhancing Moral Development

Williams and Dewett (2005) argue that a second goal of business ethics education should be the promotion of students' moral development, since the level of moral development is related to the ability to recognise ethical dilemmas, and strongly influences the perceived morality of individual choices. They draw on Lawrence Kohlberg's (1984) model, in order to explain the different levels of moral development. The first, or preconventional, level of moral development describes individuals who base their behaviour on self-interest or who blindly conform to ethical policies or codes. The second, or conventional, level describes the degree of moral development attained by most adults. Individuals on this level employ formal rules or informal societal norms when reflecting on the rightness and wrongness of actions or when considering alternatives. The third principled, or postconventional, level is attained by relatively few people, and defines the behaviour of those who make decisions based on their internalised moral standards, and who will deviate from self-interest or accepted societal rules and norms, in order to do what they believe is right. The demands made on the moral agent by both complexity thinking and deconstruction requires a high level of moral development, and in this regard, many critics may feel that the position espoused in this study may be relevant for the enlightened few, but is certainly not relevant for the average business student. However, a high level of skill and competence is needed, in order to succeed in today's competitive business world; and, if we wish for business ethics to play a vital role in the business world, the sophistication of our ethical tools should mirror the sophistication of our technical tools. Today, more than ever before, the business domain provides the forum in which human minds exercise their power, and this power should be checked by sound ethical reflection and dialogue.

In order to facilitate moral development, we should explore beyond the confines of our limited contexts and embrace an 'intersubjective understanding' (Morin 1999: 49). This is only possible if we are willing to engage in critical self-reflection. We would do well to encourage both these skills in our business ethics teaching practices, since, as Morin (44) warns, 'the worst illusions are found within intolerant, dogmatic, doctrinaire certainties'.

Morin (50) writes that there are many obstacles to intersubjective understanding, including 'egocentrism, ethnocentrism, [and] sociocentrism' – all of which constitute 'different levels of a common propensity to place oneself at the centre of the world and consider everything that is distant or foreign as secondary, insignificant or hostile'. Those who subscribe to the Cartesian dogma (in which the subject-object dichotomy prevails) are vulnerable to succumbing to these obstacles. If however, we teach the importance of openness and dialogue (and treat these as important elements when dealing with business ethics themes), as well as encourage students to imaginatively engage in situations and to be tolerant of one another, then the chances of fostering an intersubjective understanding amongst students is greatly enhanced.

The promotion of self-reflection and self-criticism is central to this teaching strategy, since '[c]ritical self-examination helps us decenter ourselves enough to recognize and judge our own egocentrism. Then we don't set ourselves up as judges of all things' (53). In other words, it is through critical self-examination that we come to understand our own weaknesses and failures, and thereby develop tolerance for other people's weaknesses and failures. An intersubjective understanding 'demands an open heart, sympathy, [and] generosity' (50), which are qualities that both reflect the level of our moral development, as well as contribute to our moral development, and which can be nurtured by a complexity and deconstructive thinking. In this regard, moral development should be understood in Rorty's (1999: 89) terms as 're-making human selves to enlarge the variety of relationships which constitute those selves'.

Morin (1999) writes that given the importance of understanding, the task of education for the future should be to develop the above qualities at all levels of education and at all ages. Learning should be focused on discussing and refuting, not on damning and excommunicating. In this regard, the art of debate and dialogue should be developed and practiced in the classroom, and it should be stressed that understanding 'neither excuses nor accuses' (52). The difference between constructive criticism and blanket condemnation should be made clear; emphasis should be placed on sources and causes of misunderstanding; and, students should be taught the skills necessary to understand across different thought structures, which, as Morin (55) notes, 'requires the ability to pass through a meta-thought structure that can understand the causes of incomprehension from one to another, and overcome it.'

In terms of business ethics teaching specifically, one could develop this quality of understanding by encouraging students to reflect on cases of misconduct, cowardice, or peer-pressure with which they can relate. Although cases like Enron and World-com grab the attention of students due to the sheer scale of corporate wrongdoing and the moral bankruptcy of key players, the average student finds it difficult to relate their own behaviour to the Andrew Fastow's or Scott Sullivan's of the world. What is far more effective, is case studies that highlight the moral dilemmas and bad situations in which ordinary, 'good' people sometimes find themselves.

Another way in which to promote understanding is to encourage students to seriously engage with the arts (especially film and literature). Although such activities fall beyond the scope of the business ethics curriculum, the arts are critical for fostering awareness of human complexity (which is necessary for understanding others), and for drawing attention to the full range of human subjectivity (which is necessary for developing compassion and sympathy). As an example, Morin argues that fictional criminals – such as the gangster kings of Shakespeare, the royal gangsters of *films noirs*, Jean Valjean and Raskolnikov – are portrayed in all their fullness in literature and film, rather than the least or worst part of themselves (as is often the case with real life criminals). Morin also uses the example of the movie tramp, Charlie Chaplin, in order to illustrate how films use psychological techniques of projection and identification, which bring us to understand and sympathise with people that we would normally find foreign or disgusting. As such, books and films help us 'to learn the greatest lesson of life: compassion and true understanding for the humiliated in their suffering' (53). In this regard, moral development is not the exclusive terrain of the learned, but first and foremost, concerns developing our understanding of what it means to be human.

Enhancing the Ability to Handle Complex Issues

Williams and Dewett (2005: 113) argue that a third goal of business ethics should be 'to promote students' ability to handle complex ethical decision making' since '[e]thical decisions involve many pieces of information to consider and multiple criteria for evaluating alternatives.' We should abandon teaching strategies that are centered on applying categorically-binding formulae or instrumental proceduralism, since such strategies merely reinforce the mistaken notion that we can 'calculate' our way out of the problems with which we are confronted today. Rather, it should be stressed that, although calculations and programmes are important (indeed, Morin (1999: 47) argues that '[w]e can use short program sequences within our strategies') they should not replace independent ethical thought, decision-making, and critical reflection – all of which are necessary for handling complex issues.

The embedded, differential model of CSR presented in the previous chapter focuses our attention on the complexity that defines our operating contexts. This complexity is, as previously argued, further heightened by globalisation, which makes it impossible to isolate a vital problem. However, as business ethicists, we can employ teaching strategies that draw attention to the challenges that arise due to our situatedness in the world. These teaching strategies should focus on developing awareness of the individual characteristics and the organisational factors that impact on decision-making, the opportunities and constraints generated by the operating environment, and the influence that globalisation plays in shaping our views of the world. Similarly, students should also be taught to reflect on the effects that their individual and corporate actions may potentially hold for these various domains.

Reflecting on the complexity of actions and consequences also requires that we acknowledge and confront the uncertainties of knowledge, and recognise that our theoretical, logical, and technical tools cannot resolve the complexity. For Morin (1999: 3), it is imperative that '[e]very person who takes on educational responsibilities must be ready to go to the forward posts of uncertainty in our times.' Uncertainty is not an excuse for inaction, nor for assuming a relativist position. Rather, uncertainty should motivate us to work harder, to engage further, and to find robust and workable solutions in the absence of meta-frameworks.

The best way in which to bring this point across is to develop teaching techniques that facilitate an active engagement in these complexities. Employing the dialectical method in class debate, encouraging story-telling, using stimulating audio-visual material, assembling diverse teams, and introducing counter-factuals to challenge commonly-accepted views are ways in which one can help students to develop a multi-perspectival, long-range view of business problems, and encourage them to become more directly involved in analysing identified problems and recognising the complexities which define our world.

A New Beginning

In one respect, business ethicists are fated to teach a codified knowledge. This is because ethics becomes institutionalised through teaching and professional practices. However, this is not a worthless exercise: it is important to remember that ethics needs morality in the same way that justice needs the law; and, that ethics should guide morality (as justice should guide the law). In other words, history is important as it provides us with moral precedents and programmes. A critical task for business ethicists concerns the choices that we make with regard to *which* precedents are taught.

Throughout this study, it has been argued that business ethicists should reject a-contextual, universal moral positions, in favour of theories and teaching strategies that focus on the complexities of the human condition, and that draw attention to the fact that learning is itself vulnerable to error and illusion (Morin 1999). As such, the type of theories that we teach should emphasise the importance of – what Morin (7) terms – the 'principle of rational uncertainty', which serves to warn us 'that if rationality does not maintain vigilance it can turn into rationalizing illusion.' Rationality should be understood as something which 'is not only theoretical, not only critical, but also self-critical' (8), as is highlighted by both complexity thinking and deconstruction. Inherent to this view of rationality is the insight that ethical action cannot be reduced to moral law. Engaging with complexity does not however lead to the rejection of 'clarity, order or determinism', but to the insight that 'we cannot program discovery, knowledge, or action' (Morin 2008: 56). Therefore, in another respect, if we – as business ethicists – take cognisance of a complex notion of ethics, we can move far beyond merely conveying a codified knowledge. This can be achieved by developing teaching strategies that sensitise students to the complexities of our world, and by equipping them with the skills needed to deal with these complexities.

For too long, we have been bathing complacently in the light of clarity and order provided by the history of moral thought. It is postulated that this fact constitutes a very prominent reason for why business ethics has had such a limited impact on the business world: business ethicists have mainly looked backwards, not forwards. However, Morin (98) argues that:

> We stand on the threshold of a new beginning. We are not in the last stages of the history of thinking, nor have we reached the limits of the human spirit. We are, rather, still in its prehistory... We are in an initial period where it is necessary to recalibrate our perspectives on knowledge and politics... And here... we must learn to work with chance and uncertainty.

The recent financial crisis revealed the cracks in the capitalistic landscape, characterised by vested interests and political power. Many would agree that now is the time to sow the seeds of new ideas, in the hope that these may take root, grow, and blossom in the cracks. The time has come to abandon programmes and solutions that have worked in the past, and to develop new strategies for the future. This study is one contribution to such a new beginning, but further research should be undertaken in order to flesh out the insights presented here. In terms of business ethics, we need to determine the exact challenge that a complex notion of ethics poses for prominent business ethics themes besides that of CSR. Research should also be undertaken in order to unpack specific teaching strategies that would equip students with the sense-making tools and tools of analysis needed to reflect on the moral dimension of business problems in their particular contexts. In terms of philosophy, we need to further develop our understanding of ethical complexity. As previously argued, deconstruction is not the only critical, reflexive, and complex philosophical position, and exploring other philosophical positions may reveal different aspects of ethical complexity, all of which can help us to understand what it means to be human.

Many people have speculated about the viability of business ethics. Having reached the conclusion to this study, it should be clear that the important question is not whether business ethics is viable, but rather how we – as business ethicists – can ensure its viability. This is the ever-renewed challenge with which we are faced.

References

Bauman, Z. 1993. *Postmodern ethics*. Oxford: Blackwell.

Friedman, M. 1962. *Capitalism and freedom*. Chicago: University of Chicago Press.

Kohlberg, L. 1984. *The psychology of moral development: The nature and validity of moral stages*. San Francisco: Harper & Row.

Lissack, M.R., and H. Letiche. 2002. Complexity, emergence, resilience, and coherence: Gaining perspectives on organizations and their study. *Emergence* 4(3): 72–94.

Montuori, A. 1996. The art of transformation: Jazz as a metaphor for education. *Holistic Education Review* 9(4): 57–62.

Morin, E. 1999. *Seven complex lessons in education for the future*, trans. N. Poller. Paris: UNESCO Publication.

Morin, E. 2008. *On complexity*, trans. S.M. Kelly. Cresskill: Hampton Press.

Rorty, R. 1999. *Philosophy and social hope*. London: Penguin.

Smith, A. 1776. *An inquiry into the nature and cause of the wealth of nations*. London: W. Strahan and T. Cadell.

Streufert, S., and G.Y. Nogami. 1989. Cognitive style and complexity: Implications for I/O psychology. In *International review of industrial and organizational psychology*, ed. C. Cooper and I. Robertson, 93–143. Chichester: Wiley.

Swenson, D.X., and D. Rigoni. 1999. Ethical problem solving and systems theory: The complexity connection. *Systemic Practice and Action Research* 12(6): 573–584.

Trevino, L.K. 1986. Ethical decision making in organizations: A person-situation interactionist model. *Academy of Management Review* 11(3): 601–617.

Trevino, L.K., and M.E. Brown. 2004. Managing to be ethical: Debunking five business ethics myths. *The Academy of Management Executive* 18(2): 69–81.

Williams, S.D., and T. Dewett. 2005. Yes, you can teach business ethics: A review and research agenda. *Journal of Leadership & Organizational Studies* 12(2): 109–120.

Index

M. Woermann, *On the (Im)Possibility of Business Ethics*, Issues in Business Ethics 37, DOI 10.1007/978-94-007-5131-6, © Springer Science+Business Media Dordrecht 2013

D

MIX
Papier aus verantwortungsvollen Quellen
Paper from responsible sources
FSC® C105338

If you have any concerns about our products,
you can contact us on
ProductSafety@springernature.com

In case Publisher is established outside the EU,
the EU authorized representative is:
Springer Nature Customer Service Center GmbH
Europaplatz 3, 69115 Heidelberg, Germany

Printed by Libri Plureos GmbH
in Hamburg, Germany